Keeping Ducks and Geese

A PRACTICAL GUIDE

DEBBIE KINGSLEY

Keeping Ducks and Geese

A PRACTICAL GUIDE

DEBBIE KINGSLEY

THE CROWOOD PRESS

First published in 2021 by
The Crowood Press Ltd
Ramsbury, Marlborough
Wiltshire SN8 2HR

enquiries@crowood.com

www.crowood.com

British Library Cataloguing-in-Publication Data
A catalogue record for this book is available from the British Library.

ISBN 978 1 78500 961 7

Designed and typeset by Guy Croton Publishing Services, West Malling, Kent

Cover design by Maggie Mellett

Printed and bound in India by Replika Press Pvt Ltd

Contents

Acknowledgements

Writing this book has been even more fun than I expected. It gave me an excuse for contacting people from all over the world during a global pandemic, and they have been generous in giving of their time, and sharing their waterfowl knowledge, experience and photos. The images received have given me real pleasure, when lockdown meant I couldn't get out and visit other flocks to take as many of my own photographs as anticipated. A particularly big thank-you is due to my husband Andrew Hubbard. To all of those listed below, thank you for the time and advice you gave so freely.

Rona Amiss
Tom Balchin, Dorset Police Rural Crime Team
Tom Blunt, Field Officer, Rare Breeds Survival Trust
Maggie Boswell

Geoff Chase
Josh and Abi Heyneke, Parc Carreg
Rose Hubbard
Morag Jones and Kate Elkington, British Waterfowl Association
Chris Just BVSc MRCVS
Ros King, The Domestic Waterfowl Club of Great Britain
Charlie Mason, The Humane Slaughter Association
Barry Nicolle
Tanya and Roger Olver, Terras Farm
Poopost Worm Counts
Claire Shand, Westgate Labs
Harriet Smith, Crediton Milling Company Ltd
Jade Stock, Out and About Poultry
Meihao Yang
Mark Wallace and Kathryn Burrell, Beaford

IMAGE CREDITS

Aaron van Cauwenberge, page 63 (top); Abi Heyneke, page 80; Andreas Trepte, page 62 (second down), 79 (top); Andrew O'Shea, page 124 (bottom); Annie Hall, page 38 (bottom), 159 (top); Barbara Griffin, page 70 (top); Barry Nicolle, page 61 (top), 63 (bottom), 64 (second down, bottom), 66 (top, third down, bottom), 67 (bottom), 70 (second down, bottom), 71 (top, third down), 73 (top), 74 (second down, third down) 78 (top), 79 (bottom), 155, 159 (bottom); Barto1666 (Wikimedia Commons), page 69 (second down); Brandon Brockett, page 73 (bottom); Brinsea Products Ltd, page 163 (top left and top right), 168, 169 (top and bottom), 172 (top); Calibas (Wikimedia Commons), page 76 (second down); Caryl Bohn, page 84; Cheshire Countryware UK/ Etsy, page 177 (bottom); Claire Peach, page 36, 70 (third down), 75 (top, second down); Clare Lovegrove, page 65 (top), 68 (top, second down, third down), 69 (third down), 72 (bottom), 154 (top); Colin Morton, page 67 (top), 74 (top), 78 (second down); Crediton

Milling Ltd, page 96 (right), 97 (top); David Iliff, page 63 (third down); Deborah Kieboom, page 60 (second down); Dick Daniels, page 69 (bottom), 71 (second down), 75 (bottom), 76 (third down), 77 (top), 78 (bottom); Elissa Butler, page 46 (left), 173; Ellie Clark/Marley Farm, page 21; Emma Scillitoe, page 10, 132; Evie Davis, page 161; Fleur Ketley, page 16, 34 (bottom), 38 (top), 44 (bottom right), 52, 58, 85 (left), 87 (right), 92, 95 (left), 99, 104, 120, 123, 176 (top); Footas van Robin (Wikimedia Commons), page 184; Geoff Chase, page 67 (third down); Heather Birnie, page 18 (top), 32 (bottom); Holly Harding-Smith, page 72 (second down); Humane Slaughter Association, page 137; Ian Gereg, page 29 (top); Ianeré Sévi (Wikimedia Commons), page 91; Icelandic Down, page 176 (bottom), 177 (top); Ikeuchi Toshio, page 67 (second down); James Ravilious/Beaford Archive, pages 12 (top and bottom), 17, 27, 37, 81, 93, 105, 121, 133, 138 (bottom), 151, 175; Jayne Hobbs, page 15; Jonathan Beilby, page 79 (second down); Karen Thorne, page 19; Kolomyia (Wikimedia Commons), page 179 (bottom); Lane Ream, page 174, 179 (top left); Lantuszka (Wikimedia Commons), page 20; Lindamacphotography, page 25 (top); Malcolm Reeves, page 60 (bottom), 61 (second down), 62 (third down); Marcus Morgan/Green Frog Designs, page 42, 43 (top); Margaret Griffin, page 25 (bottom), 87 (left), 106; Mark McCandless, page 65 (bottom); Matt Foweraker, page 61 (third down); Meihao Yang, page 195 (top and bottom), 196, 197 (top and bottom), 198 (top left, top right and bottom); Michal Klajban, page 71 (bottom); Morag and Derek Jones, page 29 (bottom); Morag Jones, page 30, 62 (bottom), 64 (third down), 77 (bottom), 78 (third down), 118; Nataliia Yanishevska, page 179 (top right); Nicola Chesshyre, page 63 (second down); Paul-Erwin Oswald, page 79 (third down); Peach Croft Farm, page 180; Peter Massas, page 74 (bottom); Red Top, page 115; Richard Bartz, page 13; Robin Monaghan, page 77 (second down); Rosemary Sharpe,

page 60 (top); Ros King, page 61 (bottom), 73 (third down); Rowan Limb, page 14; Rupert Stephenson, page 68 (bottom); Sarah Cox, page 66 (second down); Sarah's Soapery, page 181; Shutterstock, pages 85 (right), 108, 109, 186, 187; Simon Verbiest, page 18 (bottom), 26, 56, 62 (top), 69 (top), 72 (top), 73 (second down), 116; Stefhan Stjärnås, page 65 (second down); Steve Dace, page 65 (third down); Tanya Olver/Terras Farm, frontispiece and pages 32 (top), 41 (top), 44 (top), 103, 135, 136, 150, 153 (middle), 164, 182, 183; Terry Howell, page 76 (top); Thomon (Wikimedia Commons), page 75 (third down); Tony Hisgett, page 60 (third down); Tornado Wire Ltd/John Bowler Eggs Ltd, page 47 (right), 128; Tracie Hamer, page 77 (third down); trapbarn.com, page 124 (top); Wendy Anderson, page 46 (right); Wild Meat Company, page 33; Wikimedia Commons, page 178; Wolfgang Wandel, page 72 (third down); Woodstream Corporation, page 126 (top, middle and bottom)

Author: pages 8, 20 (top), 22, 24, 31, 34 (top), 39, 41 (middle and bottom), 43 (bottom), 44 (bottom left), 47 (left), 48, 49, 50 (top and bottom), 51 (top and bottom), 53, 54, 57, 64 (top), 76 (bottom), 82, 86 (left), 88 (top left, top right, bottom left, bottom right), 89, 90, 94 (top left, top right, bottom left), 95 (right), 96 (left), 97 (bottom), 98, 100 (top and bottom), 101 (left and right), 102 (top and bottom), 107, 110, 111 (left and right), 112 (top, middle and bottom), 114, 119, 122, 127, 130, 131, 132, 134, 138 (top), 139, 140, 141, 142 (top and bottom), 143 (top, middle and bottom), 144 (top, middle and bottom), 145 (top, middle and bottom), 146, 147, 152, 153 (top and bottom), 154 (bottom), 156 (top and bottom), 158, 162 (left and right), 163 (bottom left and bottom right), 165 (top left, top right and bottom), 166 (top left, top right, bottom left, bottom right), 167 (top left, top right, bottom left, bottom right), 170 (top and bottom), 171 (top and bottom), 172 (bottom), 174, 189, 190, 191 (top and bottom), 193.

Introduction

Keeping livestock can be a lifelong voyage of discovery, and no one knows it all. Ducks and geese are no different from other livestock in that respect. Happily you don't need to be an expert to bring home your first birds: if that were the case only ornithologists would be keeping them, the rest of us being too intimidated to get started. What you might need is a simple but comprehensive guide to point you in the right direction and give you the confidence to get going, and for ongoing reference as new situations and questions arise.

This book is a beginner's guide to keeping ducks and geese, enabling you to put in place the basic requirements for having happy, healthy, productive birds. It aims to reassure and inform from a practical keeper's perspective. You might be tempted to rush to the chapter on the different breeds in order to create your bird shopping list – but first be sure you have the space for them, can implement the legal requirements and create the appropriate housing, source the correct feed, plan for security, and have some basic tools and equipment at hand. That way you won't arrive home with a box of quacking, honking birds that take over your bathroom for a month before their accommodation and care are sorted.

Some readers will already have chickens, and are keen to add waterfowl to their menagerie. Some of the information included here will be familiar, but there will be plenty that is new. In my experience, waterfowl have far fewer complaints than chickens if treated appropriately, so that's some reassurance straightaway.

In these rather adversarial times, viewpoints on the best way to care for waterfowl will be as varied as any other topic. The truth is that although there are umpteen poor ways of doing things, there is also more than one way of doing things well. Use this guide as a starting point, and grow your knowledge and preferences over the years. The best way of understanding your birds is to observe them, and all you need for that is patience, time, and an interested eye.

OPPOSITE: *One of our geese taking advantage of the broad back of a heavily pregnant ewe.*

CHAPTER ONE

Us and Them: Their Place in Our Lives and Culture

There's a small corner in our seventeenth-century farmyard that has two tumbledown pigsties, and the area in front of it is – and was – known as the goose yard. It's nothing special, just a small contained area close to the house, where any kitchen peelings and scraps could be thrown to the geese and the pigs housed just behind. It was probably an area of rich smells and prodigious muck in more pragmatic times, when the expectation was that you fed your kitchen leftovers to the livestock. The corollary is that the transfer of diseases through imported foodstuffs was rather less likely in those days than it is now: four hundred years ago they spent their time worrying about flea-infested rats importing bubonic plague instead.

My favourite photographer, James Ravilious, who documented life in rural North Devon over a twenty-year period starting in the early 1970s, captured farm scenes that can seem hundreds of years old, rather than fifty. There is a reassuring timelessness about the interactions between humans and birds, and the ducks and geese portrayed could be from the 1820s or the 2020s. Some of the more ramshackle housing and fencing arrangements that he pictured can still be seen today, but it's clear that even though there might not have been the money or time to

OPPOSITE: *A bird's eye view.*

RIGHT: *The goose yard.*

George Ayre checking a sitting mother goose, Ashwell, Dolton, April 1974. Documentary photo: James Ravilious/Beaford Archive

Ashwell, Dolton, January 1977. Documentary photo: James Ravilious/Beaford Archive

spend on providing fancy bird accommodation (keeping birds was primarily for the table, whether for one's own pot or for sale to others), the ducks and geese portrayed are themselves clearly in fine fettle, plump and well cared for.

DOMESTICATION

Both ducks and geese were domesticated during the Neolithic period, about five thousand years ago, and as wild species were found in all parts of the world except for the Antarctic. The Greylag, Egyptian and Asian Swan geese are the ancestors of all our domestic geese, while domestic ducks are descended from the Mallard (originating in South Asia) and the Muscovy (from South America), with all but the domesticated Muscovy owing their parentage to the wild Mallard. Geese were first kept by the Egyptians, and in South Asia the domestication and farming of ducks spread widely across the world – which is hardly surprising since the

birds were easy and cheap to keep, and had so much to offer, from fat to feather, quills and down, as well as their eggs and the exceptional quality of their meat.

The active farming and economic value of waterfowl is reflected in the annual Goose Fairs still held in Tavistock (held in October) and Nottingham (held in October or the end of September); these fairs date back to the early twelfth century when farmers brought their geese ready for purchasers to do the final fattening for Christmas. Hundreds of geese would be driven on foot from their farms to the fairs, their feet dipped in warm tar and sand to protect them on the journey, while the feet of some were bundled into leather shoes.

Birds were certainly a desirable commodity, though it's not certain whether the local archive has so many records of people prosecuted for stealing a duck in the eighteenth century because they were starving, or because the birds had an attractive resale value.

Mallard female and male.

Greylag goose.

THEIR PLACE IN OUR CULTURE

Wild ducks and geese always stir the imagination when they fly high overhead in formation, their skeins pointing the way forwards, the laggards suggesting just how far they've travelled from their breeding grounds, reminding us of different lands and people. Ducks and geese appear in the art of cultures across the world from at least 1500BC – for example they are depicted on decorative ware made in Europe in the late Bronze Age and Iron Age. They are the subject of ribaldry, nursery rhymes, folk tales, pantomime, songs, fables, cartoons and games, and are shown (mostly dead, and draped on sumptuously laid banquet tables) in lavish still-life paintings celebrating material pleasures and the ephemeral nature of life.

They make an impact on our language, too, even in the most urban environment, the goose in particular featuring in all sorts of common idioms from 'wet goose' for someone without gumption, 'wild goose chase' (a pointless exercise), to 'what's sauce for the goose is sauce for the gander' (reciprocity), and 'having a gander' (taking a look) – while the duck has 'water off a duck's back' (having no effect), 'sitting duck' (easy target), and the even simpler 'ducking', meaning to avoid something at head height, or plunging someone under water in jest or as punishment. People whose only contact with waterfowl comes from throwing them a bagful of crusts in the park will still probably know that 'goosing' someone means grabbing or pinching them on the buttocks – although 'goosing' can mean a more generic and less physical poking or invigorating, for example of financial markets.

Ducks and geese are honoured in place names, such as Goosey in Berkshire, Polgooth (goose pool) in Cornwall, Goose Green in Pembrokeshire, Goose Craig and Goose Isle, both in Kirkcudbrightshire, while there are plenty of Duck Streets and Duck Lanes in the UK. Both birds are well represented in pub names too: The Duck Inn, The Ginger Goose,

The Dog and Duck, The Fox and Goose, The Drunken Duck, The Greedy Goose, and more.

In our local village there's an annual duck race where children use rods to catch a barrage of plastic ducks sent racing down a stream that cuts through a local farm.

Geese have been used as sport for hunting since prehistoric times, and their hunting has given rise to a form of folk art: the creation of the decoy. Antique and contemporary decoys can be sold for many thousands of pounds (and dollars – they are very popular in America). Many decoy carvers are also hunters, and the craft continues today.

Of course, a critical part of any culture is its food, and the extra effort made in its preparation at times of celebration. The feast of St Michael and All Angels is traditionally celebrated as Michaelmas on 29 September by feasting on a 'green' goose (*see* below), which augured prosperity for the coming year. Michaelmas was the first day of the farming year, and if a landlord was lucky, their tenants might gift them a goose when paying their quarterly rent. A green goose is the leaner, younger goose that was traditionally fed almost entirely on grass, stubble and harvest gleanings, as opposed to the fatter Christmas goose, which would have been finished three months later on a wheat diet, and its desirable fat reserves built up as the weather cooled.

PETS

In the thirty years that we've kept ducks and geese there has been a definite move away from small-scale keepers who saw domestic waterfowl solely as providers of meat and eggs. In times past, many will have had an older, non-productive bird kept on in a retired capacity, seen as an old friend who could teach the youngsters how things worked – but the majority of the flock was kept as a delicious

Mallards in the house.

contribution to the table. This practice does, of course, continue, but there are increasing numbers of duck and goose keepers who regard their birds as pets, a sign of more affluent times, and reflective of a culture where it is no longer expected that you have to rear your own food to assuage hunger. We may be able to source every possible type of foodstuff from a shop these days, but the yearning for getting back to nature is ever strong, and enjoying the company and shenanigans of ducks and geese can be a fulfilling part of achieving that.

CHAPTER TWO

Why Keep Ducks and Geese?

Unless you are entirely driven by commercial need, why you should keep any type or breed of livestock is very much a personal matter. One's predilections with regard to looks, the space and environment you have to offer, the taste of the meat (and in the case of ducks and geese, the eggs too), and the more intangible individual pleasure of preferring a small neat call duck or a majestic ponderous Toulouse goose, all come into play. There are further considerations, too. Overrun with slugs and snails? Ducks will be happy to eat them for you, although they may also decimate the plants you're aiming to rid of guzzling gastropods. Keen to create a thick, even grass sward in a paddock, or deal with an excess of windfall fruit in an orchard? Geese could be your new best friends, as long as you protect the trees from their nibbling bills.

OPPOSITE: *Geese are into everything.*

ABOVE: *Millhams, Dolton, September 1981. Documentary photo: James Ravilious/Beaford Archive*

Ducks and geese.

THE RURAL SCENE

Don't underestimate the pull of charm, tradition and the appreciation of the rural idyll. Having a few ducks or geese roaming your front yard, picking up worms, snails and pulling at grass and weeds, can add appreciably to your sense of wellbeing every time you look out of the window, arrive home, or spend time in your garden. It's entirely possible to have ducks in an urban back garden, too, though do consider noise issues, and avoid very vocal birds such as call ducks.

Cleverer Than You Might Think

Ducks and geese might have a brain the size of a walnut, but they are surprisingly bright and have a very good memory. We frequently change where our ducks spend the day – front garden, back garden, permanent pen, and so on. When the ducks are let out in the morning all you have to do is usher them determinedly in the new direction and they clearly think 'Oh, not like yesterday, but we remember that place, we were there six months ago, off we go!' The geese see me coming into their field in the evening, and with no ushering at all make their way towards me and their hut, to be put to bed for the night.

Black East Indian ducks.

Chickens and turkeys are a lot more awkward in this respect than waterfowl.

THE IMPACT OF DUCKS AND GEESE ON THE LAND

Ducks and geese are not interchangeable birds: geese are not simply large ducks, and their behaviour, personality, egg production, and impact on the land, and more, are very different. Historically sheep are said to be 'golden-footed', improving the ground wherever they walk, in particular with muck – although a re-wilding exponent might consider that they denude the landscape. But geese really do create an extraordinarily thick, neat sward, mucking copiously as they go, cropping the grass to an even height and creating grassland that looks more like a lawn than a paddock, although in a fastidious mood you may not want to stroll on it barefoot.

Geese have been used to keep the grass and weeds down in orchards and vineyards in many parts of the world, but their bills are the equivalent of a toddler's fingers: into everything, and quite powerfully destructive if allowed access to anything you want to keep in one piece, such as specimen trees, saplings, the electrics of a livestock trailer, the underneath of your car, or roofing felt on a poultry hut. Protect anything and everything that you value, and take care to keep it out of a goose's reach, including tree guards and fencing: removing the 'don't-want-it-nibbled' object out of harm's way is more effective than hope and crossed fingers. If you are trying to grow a wildflower meadow, they probably aren't the grazers you need for that area until the hay has been cut, baled, and taken off the field.

Ducks have very different sward habits. Ducks dibble energetically when the ground is wet. They make holes in the ground with determined bills, and continue working at it until there is no more grass and plenty of mud. If you can move them, the grass will recover well – it has, after all, also been generously fertilized with duck poo – but ducks don't so much create a lawn as damage it. To be fair, in dry weather ducks are well behaved on grass, but during and after rainfall the prospect of sifting wet mud through their bills drives them into a happy frenzy.

PRODUCTIVITY

The Golden Egg

From a completely subjective perspective, I think the duck egg reigns supreme when it comes to the egg gastronomy stakes, head and shoulders above the humble hen's egg. The first duck eggs of the season mean a celebratory breakfast, whether boiled, fried or poached. Goose eggs tend to be not so popular – the sheer size of them can be rather off-putting unless you have a lunchtime omelette, frittata, or a dish of scrambled eggs

Eggs Benedict.

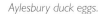
Aylesbury duck eggs.

*Decorated goose eggs
in the Pysanka Museum,
Ukraine.*

for the multitude in mind; however, plenty of people disagree, and see the goose egg as the ultimate breakfast treat.

Contrary to popular imaginings, duck eggs are not only useful for baking, but can and should be used in exactly the same way as a hen's egg if cooked for a slightly longer time because of their size. The old view is that they taste fishy, and so need to be baked in a disguising mix of flour, sugar and cake flavourings. However, this is not the case, and any fishy taste in times past was due to the ubiquitous fishmeal in their diet – and this is something you can easily avoid when choosing your poultry feed.

If you're into delicate crafts, goose eggs provide a wonderful structure for decorating. Be inspired by Fabergé and more contemporary egg carvers, painters and enamellers.

Geese and Ducks as Table Birds

Geese and ducks are supreme table birds: they are 'special occasion' poultry, particularly goose, which is the Christmas dish of tradition, well before turkeys were imported from America. Wonderful birds for roasting, both birds produce fabulous fat for making the perfect accompaniment of the best, crispest roast potatoes. As delicious as a roast bird is, there are plenty of other possibilities too: *see* Chapter 14 for recipes.

AS PETS AND GUARDS

Geese and ducks can both make friendly pets if reared by hand from a very young age, preferably from, or soon after hatching. Known as imprinting, if handled regularly and gently when small, they will recognize and trust their handler, starting with their particular human, and are normally happy to transfer that devotion to a new owner. They will contentedly sit on you, come to you, and follow you around. This is particularly rewarding with ducks, because as a species they can be somewhat anxious, unless you can spend time with them and acclimatize them to your presence.

The more you handle your birds, the tamer they will be. If you are interested in showing

Call ducks.

Young Children and Geese Don't Mix

A word of warning. The friendliest of goslings, both male and female, become protective and potentially aggressive during the mating season once they reach breeding age. If you have young children, geese are only suitable if you can keep them apart from your toddler; a goose in protective mood will break your skin with its bill, even through a thick pair of jeans.

your birds there will be plenty of hands-on activity, so they will inevitably become quiet and used to handling. A question that is frequently asked is, 'Which goose breed is the calmest and friendliest?' I would suggest that it's the handling, rather than the breed, that creates a pet out of your livestock, whether it's a goose or a cow.

On one occasion I had a lone gosling hatch, and kept it in a crate next to my desk. I held it regularly, and it spent its time nibbling my

Protective geese.

Pros and Cons of Ducks and Geese

	Ducks	Geese
Feed	Depending on breed, can be greedy	Grass makes up the majority of their diet; this needs supplementing in winter
Gender identification (if vent sexing is not an option)	Can be done by voice from around seven weeks	Can take up to six months to identify gender by behaviour, unless an autosexing breed
Suitable around young children	Yes	No
Impact on land	Negative in wet weather – and lots of poo Keep slugs and snails down	Positive – and plenty of poo but spread over a wider area
Easy to source	Yes, as hatching eggs or birds, although you may have to be flexible on breed	Fairly easy, as hatching eggs or birds, although you may have to be flexible on breed
Eggs	Depending on breed, can lay up to 300 eggs a year	Lay 30–40 eggs in the spring
Size	Plenty of choice from light to heavy breeds	Plenty of choice from light to heavy breeds
Meat	Various meat bird breeds available. The smaller varieties are not really worth the effort of plucking, but would still taste wonderful	All geese are suitable for meat (some have a bigger carcass than others). Tastes delicious. Can be a cheap way of rearing costly meat
Housing	Need secure housing	Need secure housing
Fox-proof	No	No
Lifespan	Eight to twelve years, although six years is more usual, particularly for larger breeds	Up to twenty years or more, although twelve to fifteen is more usual
Maternal/paternal traits	Some ducks never seem to go broody – how they survive is a mystery. Domestic drakes are mainly uninterested in their young	Geese are mostly good brooders. Ganders are excellent parents
Noise levels	Varies significantly by breed. All are noisy when mating and when disturbed. Some are noisy much of the time	Quiet much of the day, but noisy when let out of their hut and when they encounter people/threats
Pond	Much enjoyed but not essential; provide if you can, particularly to facilitate breeding in heavy breeds	Much enjoyed but not essential, although ideal for the breeding season; provide if you can
Space requirements	Estimate at twelve to twenty large-breed ducks per half acre if living on that ground permanently. If you have large ponds or lakes, figures of fifteen to twenty-five ducks per acre of water are normally given	Eight to sixteen adults to the acre (twenty to forty per hectare). Can be higher numbers depending on quality and density of the sward and size of the breed
Weather	Don't mind the wet or cold, but not keen on ice/snow	Don't mind the wet or cold, but not keen on ice/snow
Guards	Ineffective	Alerts the owner to disturbance/visitors
Entertainment value (entirely subjective)	High	Moderate
Keep with other poultry?	Ducks can be kept with chickens	Geese should be kept on their own patch of land

Gertie gosling.

hair, my earring, my clothes and my neck – a habit that became less enjoyable as it grew – and was altogether most affectionate. However, it proved impossible to put him back in with the flock, who bullied him mercilessly, even after many weeks of 'safe' introduction where they were kept side by side separated only by wire fencing. There can be a happier outcome, but don't expect to be able to integrate birds that have been reared separately, whether male or female, every time.

Birds treated as livestock rather than as pets are unlikely to lose their innate wariness of their keepers: it's all part of the evolutionary survival instinct. This brings us on to guarding. Geese are traditionally known as 'guard' creatures. Spend a little time with a flock and they will approach strangers and their owners alike with a hiss, necks stretched out in warning. The boldest of farmers, happy to handle a ton of bull, can be wary of a goose on a mission, and my large dogs are sensibly circumspect where our geese are concerned, giving them a wide berth when required to walk past them.

However, although any goose can nip you painfully – I've been bitten by a gander just once in thirty years of goose keeping, but once was enough! – their role as guard is predominantly about making plenty of noise as people approach, which alerts the owner to something being amiss.

SUPPORTING RARE AND NATIVE BREEDS

There are sixteen breeds of duck (of which six are a priority breed) and eleven of goose (all of which have priority status) on the Rare Breeds Survival Trust watchlist, giving you many options if you want to play a part in supporting our more endangered breeds. (*See* Chapter 6 for more detail on breeds.)

Going for the Exotic

There are many beautiful ornamental and exotic ducks and geese available (*see* Chapter 3 about the relevant legal requirements) as long as they have been bred in captivity and

ABOVE: *Shetland geese – a priority rare breed.*

Appleyard duck in a laundry basket.

have not been taken from the wild. There are various whistling, diving, dabbling, sea and perching ducks, stifftails, teal, shelducks and more, plus a whole raft of wild geese, from the Emperor to the Snow Goose, that attract enthusiasts. Many have wonderful coloration, unusual bill and body shapes, and they require additional knowledge as to their preferred habitats, breeding and feeding behaviours. It is prohibited to keep the North American Ruddy Duck in Europe in order to protect the native White-headed Duck, and the Egyptian Goose because of its ability to hybridize with other goose and duck species and out-compete native fauna for food and nesting sites.

CHAPTER THREE

Legal Matters

If taking on some ducks or geese is your first animal venture beyond caring for a cat, a dog or a couple of goldfish, don't be daunted by the fact that you are about to become a livestock keeper. Because they are considered food-producing animals there are some legal requirements to follow, even if you categorize your birds as pets. The legalities will take just minutes of your time across the year – looking after your waterfowl will take a whole lot longer than that!

If you are keeping, or intend to keep, fifty or more birds on your premises, the first step is to register the land on which you keep them, and the second step is to register as a poultry keeper. It's surprising how quickly you can accumulate fifty birds – a dozen hens, the same of ducks, a handful of geese, a few turkeys being reared for Christmas, and several hatches of ducklings, goslings and chicks can soon add up to that total.

FIRST STEP: REGISTERING THE LAND

Registering your land where you will keep your birds (even if it's a large back garden suitable for hosting those fifty or more birds) means acquiring a County Parish Holding (CPH) number for the land where the livestock will be kept. There is no charge for this. If you keep pigs or cattle, sheep or goats you will already have a CPH. The CPH is a unique nine-digit number: the first two digits relate to the county, the next three relate to

Millhams, Dolton, April 1982. Documentary photo: James Ravilious/Beaford Archive

the parish, and the last four digits are a number unique to the keeper – for example, 12/345/6789 – and its main purpose is to identify and trace the location of livestock.

To apply for your CPH, contact the Rural Payments Agency in England, the Rural

Mandarin drake.

Payments Wales in Wales, and the Rural Payments and Services in Scotland (there are no CPH numbers required in Ireland or Northern Ireland). These details will all be found on www.gov.uk. If you keep your waterfowl on someone else's land you still need a CPH if you are responsible for the birds. The CPH will link your record to the birds and the land where you keep them.

SECOND STEP: REGISTERING AS A POULTRY KEEPER

The next step is to register as a poultry keeper, which is also free. The UK government considers that you are a poultry keeper if you are in charge of the day-to-day care of the following: chickens, turkeys, ducks, geese, guinea fowl, quail, partridges, pheasants or pigeons, which includes poultry you keep as pets. You are required by law to register with the Animal and Plant Health Agency (APHA) as a poultry keeper within a month of keeping fifty or more birds on your premises. You need to complete and submit the 'Compulsory Poultry Registration Form – Keeper of 50 or More Birds', which is available on www.gov.uk.

If you keep fewer than fifty birds you can complete a simpler voluntary registration form. APHA encourages you to register as they will then be able to contact you if there is a disease outbreak (such as avian flu) in your area, and can advise you on what steps you need to take to protect your birds; this will also help prevent the wider spread of disease. The real benefit in voluntary registration is receiving this information from the source, rather than struggling with the mangled interpretations of what is required as found on social media or in the press.

If you intend to run a hatchery and have the capacity for a thousand eggs or more, or want to run a breeding site with a hundred or more breeding birds, you must register your premises as a hatchery or a breeding site.

REQUIREMENTS FOR KEEPING NON-NATIVE SPECIES

In the UK, a general licence enables anyone who has not been convicted of a wildlife crime to purchase any non-native waterfowl species, excepting the Ruddy Duck and Egyptian Goose. There is no paperwork involved for the seller or purchaser, as long as both follow the requirements set out in the licence. Birds sold under the licence must have been bred in captivity, which is defined as the bird coming from parents that were lawfully in captivity when the egg from which it hatched was laid. Documentary evidence of captive breeding must accompany any sale, hire, barter or exchange of birds. There is, however, no requirement for documentary evidence of captive breeding for imported birds.

Most captive birds sold under the terms of the licence are required to be close ringed – that is, to have at least one leg fitted with a ring that identifies the bird, and is not removable once the bird is fully grown. Microchipping is an alternative method of identification. There are many ducks and geese ordinarily resident in the UK and listed in the licence that do not need a leg ring in order to be sold.

Owners must ensure that their non-native waterfowl are not released or allowed to escape. Birds can be flight restricted by the seasonal clipping of the primary feathers on one wing as necessary, or can be kept in a covered aviary. If you cannot restrict flight or provide a covered aviary – for example with well fitted and maintained netting – then you must choose native rather than non-native species. Pinioning, where the tip of one wing is removed to restrict flight, is an option in the UK and must be carried out by a veterinary surgeon when the bird is less than ten days of age, and preferably at three to four days old. However, pinioning is defined as a mutilation, and is legally restricted in many countries.

Taiga Bean geese.

Putting on a leg ring.

Netted area for non-native wildfowl.

KEEPING MEDICINE RECORDS

It is a legal requirement to keep a record of all veterinary medicines administered to food-producing animals, including those administered by your vet or given in feed, even if you never intend to eat your birds. The record must show the following: the name of the medicine used, the supplier, the date of purchase, the date of administering the medicine (and the end date if it is over a period), the quantity of medicine used, the identity and number of the bird(s) treated, and the withdrawal period for meat and eggs as appropriate, the batch number of the medicine, and the expiry dates. The batch number will be written on the packaging, and the withdrawal period will either be written on the packaging or included in the datasheet inside.

If you lose the accompanying datasheet from any medicine, you can find them all online at www.noahcompendium.co.uk.

There are livestock medicine record books for recording all the necessary detail available from most agricultural merchants, and electronic versions are freely available to download.

Medicine records must be kept for five years following the administration or disposal of the product. It is advisable to complete your medicine records immediately after any treatment when the information is fresh in your mind and the details on the packaging are still legible.

NOTIFIABLE DISEASES

In the UK there are two notifiable diseases affecting waterfowl: avian influenza (bird flu) and Newcastle disease. Notifiable diseases are those that you are legally obliged to report to the Animal and Plant Health Agency, even if you only suspect that an animal may be affected, and failure to report it is an offence. Notifiable diseases can be endemic, which are

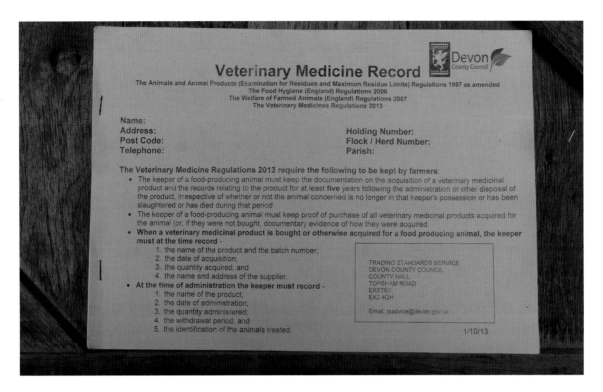

Veterinary medicine record book.

those already present in the UK, or exotic, which are those not normally present in the UK. Some endemic and exotic diseases are zoonotic, which means they can pass between animals and humans.

For most notifiable diseases there are legal powers to cull animals to prevent the spread of disease during an outbreak. Certain animals and birds, including rare breeds, may be spared from culling if this doesn't compromise controlling the disease, although this isn't guaranteed. (*See* Chapter 9 for information on diseases.)

SELLING EGGS

If you intend selling your surplus eggs direct from your farm/home/smallholding, you need to acquaint yourself with the egg trade regulations on www.gov.uk. In brief, the small-scale bird keeper wanting to sell a few surplus eggs does not need to register as an egg producer, nor to stamp their eggs with a producer code, if:

- the eggs are sold directly to consumers, are for their own use, are sold from their own home, or door to door in the local area;
- if they have fewer than fifty birds and sell at a local public market.

Note that you can only sell directly to consumers for their own use; you cannot sell to bed and breakfast businesses, cafés, pubs or shops. If you have your own bed and breakfast you can use your own eggs for making guests' breakfasts, but they need to be made aware that the eggs have not been graded (a process that registered egg producers have to comply with), so they can choose to have them thoroughly cooked if preferred. The reality is that your guests will probably be delighted to receive the freshest possible eggs for their breakfast and in their packed lunches.

If you are selling at local public markets, to ensure traceability you must have the following information on display: your name, address,

Farm-gate duck-egg stall.

Helpful Tips for Selling Eggs

The following are things not to do:

- Do not grade and label your eggs in terms of size – although you may wish to box similar sizes together.
- Avoid using the terms 'free range' or 'organic' to describe your eggs, as these are industry standards with specific compliance requirements.
- Do not wash your eggs, as this removes the protective layer provided by the bird. Keep the mucky eggs for your own use, and sell the clean ones.
- Never sell cracked eggs.

Be sure to observe the following practices:

- Be sure to know the age of the eggs you are selling so that you can be sure of the 'best before' date.
- Keep a record of your egg income, as it is taxable.
- Sell your eggs in new, clean egg boxes or trays.
- Keep the eggs you sell in good condition: it is important to store them at a cool, constant temperature, and to keep them dry and out of direct sunlight.
- Sell them within twenty-one days of laying, which in practice means they must be sold at least seven days prior to the twenty-eight day 'best before' date.

Boxed eggs.

the 'best before' date (a maximum of twenty-eight days after laying), and advice to keep eggs chilled after purchase. Personally I wouldn't dream of keeping eggs in the fridge unless I'm about to poach one, but if you are selling your eggs at a market you must abide by the rules.

You must also avoid using terms such as 'free range' and 'organic', as these have legal definitions; nor can you label them as small, medium, large or extra large either, as these descriptions have specific weight definitions. However, don't let this put you off selling your eggs; for purchasers who are keen to sample local eggs from a small-scale keeper, an open egg box showing just how large or small your eggs are is a more effective marketing approach than any legal category. If you need further guidance, contact your local Trading Standards Officer.

SELLING MEAT

You need to abide by a range of regulations when selling food. Your first step is to contact your local Environmental Health Office, as you are almost certainly going to have to register as a food business, and make that contact well before contemplating selling meat to anyone other than friends and family. An Environmental Health Officer (EHO) will visit you to discuss your intentions and inspect the premises as appropriate to your proposed venture. You will then have to meet all their requirements before advertising and selling meat.

There is some confusion over what sort of enterprises need to register as a food business. If you sell small quantities (as defined by your local EHO) of primary products such as oven-ready ducks and geese, and simple cuts such as breasts and legs, direct to your customer face to face, or to a local retailer who directly supplies the final consumer, then you might not have to register. However, if you sell anything processed (sausages, salamis, pies and suchlike), you must register. No matter what goose or duck meat product you sell, you must register if you do

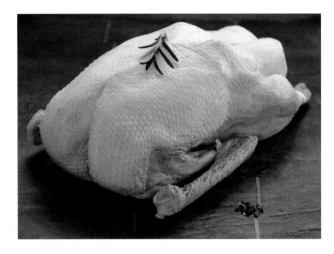

Oven-ready Legarth goose.

what is called 'selling at a distance': this is defined as anything other than selling face to face, and includes advertising your meat on the internet, whether it be on your own website or via social media, or selling your goods by phone. And you must register if your meat is transported or sent by courier to your customers, no matter how small the quantities involved.

In all cases get in touch with your local EHO so they can determine whether you really are exempt from registering or not. The good news is that registering is both quick and free.

Only if an animal has been slaughtered at an approved slaughterhouse is it lawful to place the meat on the market, including offering it for sale in your own farm shop, or serving it to guests of your bed-and-breakfast enterprise. There are poultry slaughterhouses that take what is known as private kills, accepting livestock for slaughter that goes back to the keeper in an oven-ready state, either for personal consumption or for onward sale. Before deciding to set up a goose or duck meat business, do check that there is a poultry slaughterhouse in your area.

Depending on the size of your poultry operation, you may have to become licensed to do this yourself; be aware that although this is possible, it is neither simple nor cheap. However, there is an exemption allowing poultry to be

Ducks awaiting plucking.

Geese eating weeds from the vegetable garden.

sold if slaughtered on your holding, although anyone carrying out slaughter operations for poultry that is for sale or use beyond the immediate family must hold a Certificate of Competence. For full details on this exemption please refer to Chapter 11 in the section on setting up a small poultry abattoir.

POULTRY FOR HOME CONSUMPTION

You don't need a licence or certificate of competence to kill geese or ducks to eat at home, provided:

• you own the animal;
• you kill it on your property;
• you are killing it for you or your immediate family – who live on your property – to eat.

If you kill an animal to eat you need to know how to restrain, stun and kill it humanely and quickly, without causing the animal any avoidable pain, distress or suffering. To ensure that you understand the rules and requirements for legally slaughtering your own poultry to eat at home, visit www.gov.uk (search for 'slaughter for home consumption'); and for a thoroughly detailed explanation of the process, accompanied by helpful diagrams and photos, the Humane Slaughter Association's *Practical Slaughter of Poultry Guide for the Smallholder and Small Scale Producer* is essential reading. This is critical knowledge, not only for food production, but for dealing with the emergency dispatch of sick or injured birds. More information on raising birds for meat is given in Chapter 11.

FEEDING RESTRICTIONS

It may be all too tempting to fling your leftover ham sandwich or bowl of pasta out of the kitchen window to enjoy watching your ducks frenziedly demolish your offerings, but the general rule in the UK is that once any food has been in a kitchen, whether domestic, professional or industrial, it is illegal for it to be fed to livestock – and this extends to pet poultry, or poultry not intended for meat. This doesn't just refer to foods of animal origin, but includes the vegetable portion too, due to the risk of cross contamination.

As with human food poisoning, work surfaces, hands, utensils, taps, chopping boards and more become contaminated with micro-organisms, which may then be transferred to other foodstuffs. Bacteria such as salmonella or campylobacter transfer easily, for example from raw poultry to salad, causing potentially serious human illness, and the risk is the same for your livestock. Due to the risk of cross contamination the ban on feeding kitchen waste to livestock includes vegetarian kitchens, where products of animal origin such as milk are used in food preparation. Only if you live in a fully vegan household may you feed your kitchen scraps to pet poultry.

This doesn't mean that you can't give human-related food treats to your birds. If you have a vegetable garden they will love bolted lettuce, the large leaves at the base of the plant that a human would reject, corn cobs, cabbage, and all sorts of other crops, plus the weeds you've patiently hoed. All you need do is take these straight from the garden to your birds, without going through the kitchen first.

PLANNING PERMISSION FOR BIRD HOUSING

Planning permission for small-scale bird housing is not usually necessary, and 'permitted development rights' exist for erecting structures in certain situations. Simple structures such as post and netting won't require permission, and nor would temporary, mobile, or other structures used for the purposes of agriculture. However, regulations will differ between commercial and hobby poultry keepers, so check with your local planning authority before you start any building work.

CHAPTER FOUR

Land and Accommodation

The beauty of keeping ducks is that you can do it in a small way in a fairly compact area. 'Compact' is clearly a relative term, though, and we're not talking about a couple of square metres. There is a strip of walled garden at the front of our house, three metres deep and fifteen metres long, and this is home to Hard Hattie the tortoise, a few Indian Runner ducks, the occasional group of growing youngsters if I have filled all the cage runs, and the odd waif and stray. The tortoise, the Runners and an ancient Cayuga are the only permanent inhabitants, and the ducks peer in through the windows, keeping us and themselves entertained. Geese, on the other hand, need grazing space, but this doesn't mean you can't keep a light-breed pair in a reasonably sized back garden as long as there is enough fresh grass for their needs.

OPPOSITE: *White Campbell ducklings and run.*

ABOVE: *Millhams, Dolton, December 1977. Documentary photo: James Ravilious/Beaford Archive*

HOW MUCH SPACE DO YOU NEED?

The smaller and lighter the birds, the less space you need, both in terms of their housing and their daytime outdoor areas. A trio of call ducks will fit in a pocket-sized garden and a doll-sized duck hut, whereas a pair of Embden geese need a tall, roomy shed and plenty of space to graze and roam. As with any type of livestock, start with small numbers in order to assess how the land responds across the seasons. It is much easier to add to your flock if you've initially underestimated suitable stocking rates, than to dispose of surplus birds to which you've become attached.

Suggestions for stocking rates are given below, but what is crucial is that you assess and manage the impact your birds have on your ground, and if you end up with nothing but a mud bath rather than a patch of green and pleasant land, you have too many birds on too small an area. However, don't confuse overstocking

Geese in a walled garden.

Miniature Silver Appleyard ducks.

Geese sharing their field with a ram.

with occasional periods of torrential rain, when waterfowl can create havoc with the ground in a short space of time. In very wet conditions, restrict the birds to a sacrificial area that can then be rested and allowed to regrow or can be reseeded. Creating the space to rotate your birds will be good for your land and is better for their health.

Stocking Rates

Figures of fifteen to twenty-five ducks per acre of water are normally accepted as reasonable, but as most keepers are more interested in the amount of land rather than water needed for ducks, a fairly generous estimate is twelve to twenty large-breed ducks per half acre. This equates to six to ten large-breed ducks on a patch 20 x 20m if they are living on that ground permanently. The best plan is to divide the space

so that the birds can spend a few weeks on one area, before moving to the other.

Eight to sixteen adult geese to the acre (the equivalent of twenty to forty geese per hectare) is considered appropriate, and this number can be higher depending on the quality and density of the sward available for grazing, and the size of your birds.

Take these guidelines as a starting point, and be generous rather than mean in terms of the space you allocate to your birds; they will thank you for it.

Male-to-Female Ratios

Why are male-to-female ratios relevant in a chapter on land and accommodation? Because if you keep birds of both sexes, you may need more room than you think in order to keep groups separate, particularly if you find yourself with a surplus of males during the breeding season.

Putting too many males with female birds can cause real suffering and mutilation for the females. Drakes are sexually rampant, and can damage or even kill females with an excess of ardour, so err on the side of caution. Generally accepted guidelines indicate one drake to every four to six females, but we tend to have no more than one male for six ducks, and a drake can cope with twice that number, testosterone-fuelled as they are. You can keep a group of drakes together very happily, but do not be tempted to put a female in with them even for a brief period – it isn't kind or safe. If you don't want to breed you don't need a male, but ducks do seem to appreciate a drake in their group, and if there isn't one, a female will start to mimic male behaviour, and in some circumstances can even change gender.

Ganders are keen breeders but are a little less sex obsessed than drakes. You can keep geese in pairs, or a gander will happily be in a group with four or more females, although keep the ratios down if you want him to mate with all of his females. Ganders definitely have favourites in their harem, as you will be able to monitor by the greater amount of head feathers removed from the preferred goose as a result of multiple matings.

INDOOR AND OUTDOOR FACILITIES

What infrastructure do ducks and geese need to keep them happy, healthy, thriving and safe? And what does the duck and goose keeper need in terms of facilities for their birds, which give the human part of the equation maximum pleasure and minimum grief? The first is good, solid housing so that you don't even have to think about your birds once they've been put to bed for the night. The second is determining your mindset regarding risk, which is rather like contemplating whether you're prepared to play the stock market or prefer to keep your savings, meagre or otherwise, in a savings account.

The high-risk game is allowing your birds to truly free range during the day; it may happen on day one or years later, but you will inevitably lose some, or even all, if you are unlucky, to foxes or stray dogs. Medium risk is creating a fox-proof area that your birds can enjoy while you're not around, and letting them roam more widely if you are working or relaxing in the same area. For the keen gardeners among you, this suggests having birds close to your vegetable patch, where you can keep an eye on them when they range, and you won't have far to go to barrow their precious muck on to your compost heap.

Low risk is keeping your birds contained within a securely fenced area at all times, and this means having multiple areas so that birds can be rotated and the ground doesn't become sick and riddled with parasites. (*See* Chapter 10 on keeping your flock secure.)

HOUSING OPTIONS

Please ignore ancient wisdoms that state that geese are happiest sleeping in the open overnight, and that a couple of straw bales placed against the prevailing wind are all they need to keep them protected. You may not have seen foxes or other predators, but if you have birds, they will undoubtedly come (and are quite probably there already). Geese are creatures of habit and will head towards their comfortably bedded and secure hut as dusk approaches, even if they free range all day; don't leave it until dark to shut them away.

In theory, and frequently in practice, duck huts are low structures, tall enough for ducks to stand upright plus a little more for ventilation. Goose huts are, as you'd expect, somewhat taller, to allow the same freedom of movement for the birds when they are secure for the night. In most cases these huts are not tall enough to accommodate a person standing upright.

Duck house.

Duck hut big enough for humans and a wheelbarrow.

Duck hut with monopitch roof suitable for young birds.

However, I am not happy about cleaning out diminutive huts on my hands and knees, while bumping my head on the roof. Neither do I want to have to wear a boilersuit over my clothes on sunny days in order to avoid getting grubby when collecting eggs in an inaccessible coop.

Years of trying to make our lives easier means coming to two conclusions: either have huts big enough for the keeper to stand in and wield a muck fork, with a door wide enough to accommodate the back end of a wheelbarrow, or have a monopitch roof on a low hut that you can prop open with a safe stay, and muck out that way. The former we use for adult birds, and the latter works excellently with attached cage runs for youngsters.

Housing can be charming, posh, decorative, workmanlike, cobbled together from wooden pallets and bits of spare ironmongery and timber, or a DIY sensation and costing from nothing much to a fortune. It can be purpose made, or repurposed from any sort of small building such as a garden shed, stable, pig ark, goat hut, dog kennel or lean-to.

If you want new, there's nothing wrong with buying a garden shed for your birds – it will be cheaper than a new purpose-made bird hut, tall enough for you to move around inside, and with a few minor adjustments will make a great duck or goose home. Helpful adjustments would usefully include lining the base with weldmesh to deter rats, and removing glass windows and replacing them with acrylic sheet and/or weldmesh, depending on the prevailing weather and the amount of ventilation needed.

Most important is that you choose something robust that a visiting fox or badger can't chew through overnight, and that won't be blown over or apart in a storm. The easier it is to keep clean the better, and ventilation is critical, but avoid creating through draughts.

How much housing space do your birds need? Large adult ducks require around 0.4sq m (4sq ft) per bird in the hut, if they are also able to roam freely during the day. Small ducks such as calls can cope with half this space. Adult geese will need about 1sq m (10sq ft) each.

Recycled plastic duck or goose hut.

*Recycled plastic animal
ark suitable for geese.*

Materials for housing include marine ply, shiplap, timber planks, recycled plastic and more. In damp areas – which is most of the UK – avoid softwoods if you want a long-lasting hut, although they will do the job for a few years. Recycled plastic huts are very long-lasting, easy to clean and less prone to mite infestations. If you choose roofing felt to cover the roof of your hut and geese can reach it, they will rip at it, so be prepared to replace it every year or two.

An alternative roofing material for tall or low-level huts that is waterproof and can't be chewed off would be recycled polyethylene stock board (more frequently used to sheet cattle gates); it is inexpensive and available from agricultural merchants. Corrugated tin roofs can work well, but drip with condensation in cold weather and get very hot in summer, so consider the anti-condensation type, or insulate it; fibre-cement roof sheeting is also a good option. For wooden doors with goose- or rat-chewed bases, reinforce the lower edge with a plate of galvanized tin.

It is usual for hen houses to be mobile, but unless you have small ducks that you want to

Goose-hut door with galvanized metal strip at the base of the door.

Rouen × Aylesbury meat ducks.

Using old bed sheets to create shade in hot weather.

Using a hen hut for a pair of geese; a useful temporary measure.

reposition regularly, it's more likely that your waterfowl huts will be static. To deter rats from nesting under the hut and nibbling at their leisure at the base, consider putting down a thick concrete pad as your hut base (we're talking 15cm in depth); it could be made with an additional apron large enough to stand drinking water containers, which will help avoid muddy bogs round the drinking area.

Ensure that doorways are good and wide so that your birds don't have to wait patiently in line to emerge each morning. Of course they won't wait patiently, so will tumble over one another and squeeze out uncomfortably if the opening is too narrow, potentially damaging themselves and each other. If the hut is more than a few centimetres off the ground, put in place a shallow non-slip ramp, as wide as the door, to avoid causing leg problems. We use an old stable for some of our adult ducks, but its stone threshold is slightly too high for waterfowl. However, a sturdy wooden pallet is enough to give them a helpful step into their accommodation.

Create an area that throws shade to protect your birds in very hot weather – this might be an extended roof with open sides, or house the birds under the protection of trees, shrubs or hedging. Cage runs can be covered with old sheets on a hot day to create shade, or they can be moved into areas where there is natural protection. However, ensure you don't have overhanging branches that are accessible to foxes, who can use this as a way into your pens. In icy weather put a solid-topped pallet or a heap of straw on the ground so birds can get off frozen soil and snow.

Ducks and geese neither need nor want compartments or nest boxes inside their housing. Give them a good layer of bedding and they will create their own nests in which to lay.

Types of Bedding

From the tender age of a day old to maturity, ducks and geese are fairly indiscriminate in their mucking. Ducks in particular squirt out liquid poo that will go up the walls of their housing as well as on the bedding, and while geese may produce firmer muck, they make up for it in quantity.

For duckling and gosling care, including bedding options, *see* Chapter 12, but for growing and adult birds use straw, softwood shavings or shredded paper. All these bedding materials compost well and will be a great addition to the vegetable bed once well rotted. Avoid hardwood shavings as they contain tannins that are toxic for the birds. Hay is unsuitable as bedding as it can get tangled round their feet and becomes a very heavy moisture-retaining mat, beloved of bacteria and mould. Avoid sawdust as birds can eat or inhale it.

We never leave food or water in the house; overnight the birds will be asleep, and during the day it's best to put food and water in the run to limit the extent to which bedding will get wet and dirty.

ACCESS TO WATER

The field where we keep our geese has a ditch running through it, fed by water from a spring and rainwater. The geese love to swim in it, and will use it to mate in, but they certainly don't choose to visit it on a daily basis. They also hop into a sheep water trough in hot weather, and thoroughly bathe and preen themselves.

During daylight hours waterfowl need constant access to drinking water, and water to preen themselves. Preening is vital for keeping feathers in prime condition, the birds first wetting their feathers and then using their bills to realign their feathers. As a minimum, ducks and geese need to be able to fully dip their heads into water, cleaning their eyes and nostrils, but wherever possible provide a pond, too. If this is a man-made structure rather than a running stream, pond or ditch, you will need to change the water frequently, so make sure that when filled with water you can tip it out, or at least bucket it out before cleaning and refilling.

A children's paddling pool or sandpit makes an excellent duck pond.   *Mallard ducklings having their first supervised swim.*

Ducks love a pond even more than geese, and there are many simple containers that can be used if a dug-in pond isn't suitable for your set-up. Children's sandpits or paddling pools, rigid pond liners, large planters (without drainage holes, obviously), old butlers' sinks with the drain hole bunged – all of these make good small ponds that can be easily cleaned and regularly refilled. Put a shallow solid ramp in place so the ducks can access the water, and put a breeze block or similar into the water so they can rest and get out again easily. I've left a filled bucket of water for half an hour in the garden and found a duck inside it on my return, unable to get out without my help – so do be careful.

Even if you have a pond, make sure that additional clean drinking and head-dipping water is always available, changed daily, and that there is plenty of it, particularly in summer; waterfowl will foul any water they can get into, so be prepared to empty out and refill troughs and ponds regularly. This isn't simply for prettiness: anything nasty in the way of parasites or disease will spread to every bird that uses the same watering hole.

Don't allow ducklings access to swimming water when unsupervised; until they are fully feathered and are preening to distribute oil from their oil gland to make themselves waterproof, their down is easily saturated and there is a high risk that they will drown. This is particularly important for artificially incubated young who aren't picking up some oil from contact with their mother. Furthermore, don't put young ducks or ducklings in a pond where there are already adult drakes, as it's not uncommon for the drakes to drown the young ones. Wait until the youngsters are fully mature before giving them access to a shared pond – and even then be conscious that drakes can't resist mating with new blood, and their strong sex drive means that fresh young females can be held under water during mating for too long and will drown.

Rigid pond liner and feed trug.

Poultry fencing.

If you plan to excavate a pond and need to line it, remember that ducks and geese have very sharp toenails, so use the heaviest, thickest, most robust pond liner you can. In addition, deal with any sharp objects at the base and sides of the pond by removing them and lining the pond with a layer of sand or pea gravel. If you are excavating on heavy clay you can puddle the clay to create a free but labour-intensive layer that will hold water (puddling is compacting the clay to remove any air, achieved by plenty of trampling).

However you choose to line your pond, ensure gently sloping sides at several points so that birds don't struggle to get out, and lay flat stones down these access areas to avoid them becoming mudslides.

FENCING

Creating a secure fenced area (or multiple areas) for waterfowl is not a cheap business. If you are creating a permanent series of runs,

you want to build them in the expectation that they will last you fifteen years or more. This being the case, if you are planning on using livestock fencing timber, make sure you buy posts that are guaranteed for at least that long. Your fencing needs to meet your specific requirements; for example, if you want several small areas for separating breeding groups, your internal fencing will need to be almost as robust as the perimeter. But if you have just a single group of ducks to produce eggs and meat for the table, something simpler would suffice. You might also want your birds to sift through the vegetable garden come winter, in which case you'll want to fence around that area, too, to keep them safe from predators.

There are plenty of keepers who create secure pens and never shut the doors to the goose or duck houses in the evening; however, having seen the extraordinary endeavours of a hungry digging vixen that laboured in peace all night, this is not a scenario that I could live with.

Ornamental wild ducks and geese may not comply with being housed at night, so your perimeter fencing will need to be very effective, standing 2m (7ft) high, with an overhang projecting outwards from the top and an outward-facing apron of mesh dug at least 30cm (12in) below ground level. (*See* Chapter 10 on creating fencing that keeps your birds secure.)

Sagging fencing can be climbed by foxes, so keep fences taut, and consider using chain link for enclosing very large spaces. Heavy-gauge weldmesh is useful in smaller areas, and can be put along the base of a high-tensile or chain-link poultry fence to keep ducklings and goslings in, and mink, weasels and rats out. Avoid chicken wire as the main fencing material, as it is eventually breached by all chewing creatures from rats to foxes, and can't be pulled taut. Heavy-gauge 25mm chicken wire (aka hex netting) can be used as a cheaper alternative to weldmesh along the base of your main fencing to keep small creatures in (and out). If you choose a colour-coated fencing, opt for black as it is less obtrusive.

Cage or Covered Runs for Young Birds

For birds that have been hatched artificially, or for parents with newly hatched youngsters, extra protection is needed against rats, cats, buzzards, crows and more. We make simple, low-level covered runs that pull up to mobile huts, with mesh tops and sides. For the very youngest birds just out of the brooder with no watchful parent alongside, we have a run that also has mesh on the base. We use pheasant netting or chicken wire as mesh, stapled to the timber frame. Add timber skids with curved ends under the hut, or add wheels, so the hut can be pushed or pulled across the ground without too much difficulty. The huts and runs can be moved on to fresh grass every couple of days.

You will need to incorporate a hinged lid to put in feed and water, and to open and close the hut door; we make this with ply and cover it with roofing felt. When goslings are old enough to rip at this, it's time to move them into something bigger. Although the timber sits directly on the ground, we find that the runs last five or six years before needing a few running

Cage run and hut.

repairs, and twice that long before they are past repair. Cage runs are for daytime use only (and ours are visible from my office); the birds must be safely secured in their huts at night. If runs are out of sight use weldmesh rather than netting or chicken wire to keep youngsters secure.

Electric Fencing

We have used electric fencing satisfactorily for years for ducks and never had a problem with it, but we wouldn't contemplate an electrified netting for geese as they get their necks through the mesh and easily become entangled, and not surprisingly very stressed. For ducks use the tallest poultry netting, which is 1.2m (4ft) high, and be prepared to strim the grass by the base of the fence regularly to stop it shorting out. To do this, move the posts inwards a pace, strim, then put the fence back – you don't want your strimmer, shears or mower to cut through the fence.

Remember to check the fence is on and working each day – and you don't need to shock yourself to find out, as pocket-sized electric fence testers are readily available; you should also hang a sign on the fence that warns any visitors that it's electrified. Ducks that are flighty should be wing clipped to stop them flying over the fence (*see* Chapter 7).

Electric poultry fencing comes in 25m, 50m and 100m lengths, and in addition to the fence itself you will need an energizer, an earth rod and a power source: a leisure battery, solar panels or mains. Solar is a useful back-up for keeping the battery charged, but may not be reliable enough on its own in extended cloudy weather.

Electrified wires can also be used on the outside of permanent fencing, using insulators to distance the electrified wire from the main structure. To deter foxes, place one strand a few inches above ground, another about a foot up, and if you are concerned about climbing predators, another along the top. As with all electric fences, it's a legal requirement to display yellow warning signs that feature a lightning sign in a triangle, the symbol associated with electricity.

Think about where you are going to site the gate to your pen and its design, as you'll be going in and out at least twice a day, with

Ducks in the garden with electric fencing.

Electric fencing kit.

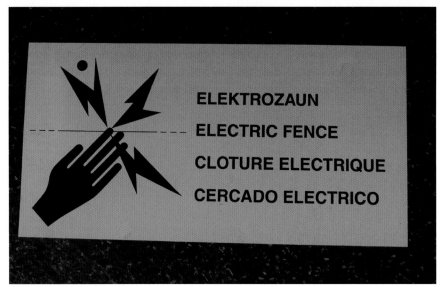

ELEKTROZAUN

ELECTRIC FENCE

CLOTURE ELECTRIQUE

CERCADO ELECTRICO

Electric fence sign.

feeders and possibly buckets of water or a hose. We lay paving slabs under the entrance, which stops predators digging under the gate and provides a solid surface in wet weather. Hinge the pen gates so they open outwards, so if some great big gander gasps its last at that spot, or if buckets get dislodged, or any other object such as a fallen branch bars the way, it doesn't prevent you from getting in.

As a little addendum about fencing, we often co-graze our geese with sheep, and have created a goose pop-hole in the fencing so that the geese can get to their hut and feed, and the sheep can't. The geese quickly work out how to use it, and we can close off the pop-hole if we need to keep the geese in a contained area for any reason.

Goose pop-hole.

*Goslings and large-breed
ducklings sharing
a cage run.*

Keeping Geese and Ducks Together

We frequently have artificially incubated goslings and ducklings hatching at the same time, and they are reared happily together to start with. If the ducks are a large breed they are fine to be kept with young geese for a few months, but without fail they are always separated well before the following spring and the onset of the mating season. Single-sex groups work best if you have large numbers of males. For smaller breeds of duck and for the vulnerably slim-necked Runners, separate them from any goslings once they move from the brooder and on to grass. You don't want young geese squashing or trampling young ducks, or using their bills too eagerly on them as they go about their business exploring everything around them.

CHAPTER FIVE

Tools and Equipment

The joy in keeping ducks and geese is that you really don't need any amount of fancy tools or equipment to get started. It's not essential to have access to a 10-ton digger to create a lake complete with island in order to enjoy waterfowl. You don't need a lake – although how wonderful for them, and you, if you do have one. This short chapter covers the essential equipment, the things that will make your life easier if you have them, and those pieces of kit that are worth asking for on your birthday. Of course, what is essential to some, will be

OPPOSITE: *Duck in a rubber trug.*

ABOVE: *Pilgrim geese by the front door.*

entirely unnecessary to others. If you intend to artificially raise birds from hatching eggs, you can't do it in an airing cupboard – but if that's not part of your plan, you can dispense with budgeting for an incubator.

I hope this information will stop you for the most part having to scour each chapter in order to make a shopping or DIY materials list, as I had to do when reading about starting out many decades ago.

UTTERLY ESSENTIAL EVERYDAY ITEMS

You may not think the items described below are essential, but all I can say is that we would be lost without them. Many items can be bought second or third hand (although I'd exclude a good avian vet from that category!), but do your homework first on prices for new kit, as some used things may be a false economy, costing almost as much as new and having a much shorter useful life. Items such as galvanized metal drinkers corrode and leak with age, which makes them entirely unfit for purpose,

no matter how cheaply you bought them at the local car boot sale. I'd also baulk at buying pre-owned poultry housing unless you are prepared to give it a very thorough disinfecting before allowing your birds anywhere near it.

It is important to note that quite a lot of this kit (in particular the hand tools) may well be in your possession already – you don't need to buy duplicate items just for your birds.

CLOTHES AND FOOTWEAR

Clothes aren't tools, but count as essential equipment in my book. As a year-round welly wearer, neoprene-lined ones for winter are key to warm feet and good moods, while unlined ones are cooler and much more bearable in hot weather. They also give a possibly over-optimistic sense of protection when heading into an area where you know there might be rats (*see* Chapter 10 for how to deal with rodents). You may prefer sturdy work boots in all but the wettest weather; in any case, you don't want to trip around in kitten heels or brogues to do your bird chores. Comfortable socks are crucial to

My birthday wheelbarrow.

Essential Everyday Items

Transporting your birds
Crate — For transporting your birds; cardboard boxes are fine for short journeys

Housing, feeding and drinking

Housing	*See* Chapter 4 for options
Drinkers	You can choose from shallow water buckets, automatic drinkers, galvanized and plastic drinkers. It's useful to have a range of options depending on the size of bird, and to have something in reserve in case of breakages
Feeders	For ducks use feeders with lids wherever possible to keep feed dry. Treadle feeders are a good option for keeping rats and wild birds at bay. Geese will eat wheat from a rubber trug or shallow bucket filled with water (which is also a rat deterrent)
Buckets	For carrying water, feed, tools, shavings, eggs and so on. Shallow buckets make excellent drinking troughs
Rodent-proof feed bins	Lidded metal dustbins or old chest freezers with the locks removed are cheap, effective options
Feed scoops	These don't need to be bought – any container that you know holds the right amount of feed will do (large yogurt tubs, empty fencing staple buckets, and so on)
Hosepipes	To get the water where you need it

Mucking out

Wheelbarrow	Plastic or metal
Muck fork	For cleaning out your bird huts
Shovel	As above

Egg storage

Egg trays	Essential for storing eggs awaiting incubation or eating. Plastic trays last a lifetime, recycled paper ones can be composted when finished with.

Other useful tools

Spade	To dig out that trench round your runs for burying fencing, and for many other purposes
Hammer	Hand, club and sledge, for fencing, making and mending housing, and all sorts of repairs and waterfowl-related DIY
Crowbar	Lots of uses, including repurposing pallets for housing
Saws	Pruning saw for keeping undergrowth in trim and carpentry saws for woodwork
Penknife	A penknife in every jacket pocket saves time and tempers (for opening feed sacks, cutting straw-bale twine, etc)
Slash hook, loppers and secateurs	For keeping poultry areas clear of scrub, nettles, etc
Fencing tools and consumables: iron bar, post-bumper, wire fencing tools (fencing pliers, strainers etc), graft, staples	Even if you pay someone to do your fencing, there will inevitably be repairs, but hopefully not for a decade or so
Cordless drill	Not just drills but cordless anything for working in places where it is difficult or impossible to run electric extension leads. An extra battery is useful so you won't have to wait hours for a recharge while you're in the middle of a job
Torch	Head torches in particular are brilliant on the hopefully rare occasions when you need to attend your birds at night or peer into dark corners

Animal care and first-aid essentials

Wormer	Flubenvet, available both loose or incorporated in feed for ease
Antiseptic spray	Known as violet or purple spray – use like a liquid plaster on minor wounds
Antibiotic spray	For wounds where you are concerned about secondary infection (vet prescribed)
Petroleum jelly	Soothing on legs, feet, etc
Strong farm detergent	Many makes available. Use to clean huts and equipment, and as foot dips for biosecurity

Other essentials

Somewhere dry to store straw/bedding	You don't want soggy bedding
A good vet	Ask around to find someone with avian experience
Veterinary medicine record book	This is a legal requirement

Bohemian goose with a shallow water bucket.

Other Useful Pieces of Kit

Reference books	*See* booklist in Further Information
Electric fencing kit (and batteries, energizer, tester, warning sign)	*See* Chapter 4
Rigid pond liner	Something that's not too heavy to tip out when full of dirty water. A redundant children's (watertight) pool or sandpit makes a good alternative
Incubator	Many types available – *see* Chapter 12
Candler	Various types available – *see* Chapter 12
Heat lamps and bulbs	*See* Chapter 12
Hatcher	*See* Chapter 12
Brooder	Home-made or bought. *See* Chapter 12
Freezer	For storing your home-reared oven-ready poultry
Blowtorch	For removing the last bits of feather and down from plucked birds
Boiler or steamer	To loosen feathers and speed up plucking
Closed rings and applicators	If keeping relevant wildfowl, and/or you intend to keep breeding records
Pest and predator prevention	*See* Chapter 10
Chainsaw and associated safety gear (helmet with visor, boots, trousers and gloves)	If you need to clear branches by poultry runs a chainsaw will save hours and muscles, but please get all the necessary training and safety gear; this is not a piece of kit for a casual approach. If you are not competent, hire a professional
Bolt cropper	For cutting through chunky weldmesh (life is too short for using a hacksaw)
Small cordless angle grinder	For cutting through chunky weldmesh

avoid blisters when you are likely to be wearing wellies or heavy boots for extended periods. And for winter keep a handful of beanie hats by the front door to keep head and ears out of rain and sharp winds.

Gloves of many kinds, from rubber-palmed builders' gloves to thick leather when handling fencing materials, are other essential items, and I particularly like disposable vinyl ones for any messy jobs, such as dealing with a wound, when you don't want to introduce bacteria. Use tough gloves, such as leather gauntlets, when handling your birds – they have sharp claws as well as soft webbed feet.

A decent pair (or two) of waterproof jackets and trousers will keep you dry on those days when there are no gaps in the rain; and for particularly mucky jobs and protection against poo, farm boilersuits can't be beaten.

Well-used gloves.

CHAPTER SIX

Duck and Goose Breeds

There are many waterfowl breeds across the world, from the stunning multicoloured King Eider to our common Mallard. The following guide focuses on those you are likely to find in the UK, and includes all the rare and native UK domestic breeds, plus the ornamental breeds commonly kept, and a selection of the wildfowl that live or make their way to the UK for part of the year. Those marked * are on the Rare Breeds Survival Trust's watchlist, and those marked ** are their priority breeds. The names given in brackets show alternative names.

Heavy breeds are most commonly seen as meat birds, but medium and light birds are equally good eating although yielding less meat (sometimes a great deal less) in exchange for the effort of plucking. Equally, the heavy meat bird will produce large, tasty eggs. Be aware that although average weights or weight ranges are given, as are numbers of eggs laid per year, every bird is individual and will have its own particularities. For the precise breed standards refer to the most current British waterfowl standards produced by the British Waterfowl Association.

OPPOSITE: *Geese in the flower garden.*

Goose breeds	Classification and type	Weight of live birds	Information
African	Domestic. Heavy (Asiatic breed) Meat, broody	Gander 10–12.7kg (22–28lb) Goose 8.2–10.9kg (18–24lb)	Large, dramatic-looking bird standing a metre tall, with colour and markings as its close relative the Chinese goose, both having a prominent knob at the top of the bill. Surprisingly gentle for its size. Colour variations are brown or grey, buff, and white. A brown stripe runs from the top of its head to the base of its neck (however, the white is pure white). It has a smooth, semi-circular gullet or dewlap that develops between 6 and 36 months of age. The meat is lean. It produces 20–30 white eggs in a season, and goes broody
American Buff	Domestic. Heavy. Meat, broody	Gander 10–12.7kg (22–28lb) Goose 9.1–11.8kg (20–26lb)	Large goose with the same markings as the Toulouse and Greylag. Rich orange-buff colour with orange bill and webs. Lays 20–30 white eggs per year, goes broody; good parents. Rarely flies. Needs access to grazing and swimming water. Docile
Barnacle	Wildfowl	1.3–2.2kg (2.9–4.4lb)	Medium-sized, sociable goose, with black head, neck and breast, with creamy-white face, white belly, blue-grey barred back and black tail. They eat leaves and the stems of grasses, roots and seeds
Brecon Buff**	Domestic. Medium. Meat, broody. Grazing management	Gander 7.3–9.1kg (16–20lb) Goose 6.3–8.2kg (14–18lb)	Deep buff throughout with pink bill, legs and webs and brown eyes. Developed from wild buff geese in the Brecon Beacons crossed with an Embden. Dual-lobed abdomen. Bred as a table bird. Good broodies and parents. Hardy birds producing 20–30 white eggs per year. Rarely fly but forage widely

Goose breeds	Classification and type	Weight of live birds	Information	
Brent	Wildfowl	1.3–1.6kg (2.9–3.5lb)	Small, dark goose with black head and neck and grey-brown back. Adults have a small white neck patch. They eat eel-grass and other vegetation	
Buff Back (and Grey Back)**	Domestic. Medium. Eggs, meat, broody	Gander 8.2–10kg (18–22lb) Goose 7.3–9.1kg (16–20lb)	**Buff:** European pied or saddleback goose. The head, upper neck, back and thighs are buff, the lower neck, breast and flanks are white. Very similar to the Pomeranian, but with dual-lobed abdomen. Blue eyes and orange bill and legs. Will lay at least 30 white eggs per year, goes broody; make good parents. Rarely flies **Grey:** Same type and pied, saddleback patterning as the Buff Back, with grey colour instead of buff	
Canada	Wildfowl	4.3–5kg (9.5–11lb)	Large goose, with distinctive black head and neck, and white throat. Introduced from North America, it has successfully spread to cover most of the UK. Eats roots, grass, leaves and seeds	
Chinese	Domestic. Light (Asiatic breed). Eggs, meat, broody. Guard	Gander 4.5–5.4kg (10–12lb) Goose 3.6–4.5kg (8–10lb)	Much smaller than its similar African cousin, and with a much larger knob, in particular in the gander, and no dewlap. Long slender neck, which they use to investigate all things. Come in brown or grey, and white variants. Heavy egg layer, producing 40–80 white eggs a year, which may be split into spring and autumn seasons. They are known as good guard geese, and are voluble and not safe around small dogs. Susceptible to cold (in particular the knob) so need shelter from very cold weather. Can fly	

(continued...)

Goose breeds	Classification and type	Weight of live birds	Information
Czech (Bohemian)	Domestic. Light. Eggs, broody	Gander 5–5.5kg (11–12lb) Goose 4–4.5kg (9–10lb)	The smallest domestic goose, lively nature, white in colour with blue eyes, orange-red bill, legs and webs. Female is shorter than the male, with thicker neck and deeper body. Conversational (a polite way of saying rather noisy) birds that make good broodies and parents. Lay 40–60 eggs per year
Egyptian	Wildfowl	1.5–2.25kg (3.3–4.9lb)	Related to the Shelduck, pale brown and grey goose with dark brown eye patches and contrasting white wing patches seen in flight. Introduced as an ornamental bird, it escaped into the wild and is now considered an invasive alien species in the UK; it is not permitted to import, keep, sell, transport, breed or release them
Embden**	Domestic. Heavy. Eggs, meat, broody	Gander 12.7–15.4kg (28–34lb) Goose 10.9–12.7kg (24–28lb)	The tallest goose, standing at over a metre, large and long. Pure white birds with light blue eyes, and orange bill, legs and webs. Their height indicates the pure Embden. A protective breed, not suitable to be around small dogs or young children, particularly at breeding time. Lay around 30 white eggs per year
Emperor	Wildfowl	Gander 2.7–3kg (6–7lb) Goose 2kg (4.4lb)	Distinctive stocky breed, resident of Alaska and north-east Siberia. Naturally tame, and popular in wildfowl collections, doing best in cool climates. Lay 2–7 creamy white eggs in early summer, with a short incubation of 24–25 days

Goose breeds	Classification and type	Weight of live birds	Information	
Flanders (Flemish)	Domestic	Gander 5–6kg (11–13lb) Goose 4–4.5kg (8.8–10lb)	Small, pied, saddleback-type goose originating from Flanders, half the size of the Pomeranian. Endangered rare breed. Hardy bird with high quality meat and quills. Lay around 60 white eggs per year. Note the lack of an asterisk: they are reserved for Rare Breed Survival Trust listed and priority breeds, and the Flemish is not a UK breed	
Franconian	Domestic. Light. Meat and feathers	Gander 5–6kg (11–13lb) Goose 4–5kg (8.8–11lb)	Small, hardy active goose from Southern Germany that raises good sized clutches of quick-growing goslings. They require little feed in harsh winters. Colours: snow white, occasionally grey, white mottled and blue, with light blue eyes. Bill pale orange-red with a pink bean. Lays around 20 eggs per year. Only recently brought to the UK	
Greylag	Wildfowl	2.9–3.7kg (6.4–8lb)	The ancestor of most domestic geese and the largest of the wild geese native to the UK and Europe. They eat grass, roots, cereal leaves and spilled grain	
Hawaiian (Nene)	Wildfowl	Gander 1.6–3kg (3.7–6.7lb) Goose 1.5–2.5kg (3.3–5.6lb)	The official bird of Hawaii. Medium-sized breed, similar colouring for male and female with black head and hindneck, buff cheeks and furrowed neck with black and white diagonal stripes. The breeding season lasts from August to April, longer than any other goose, with eggs laid between November and January	

(continued...)

Goose breeds	Classification and type	Weight of live birds	Information
Pilgrim**	Domestic. Light. Meat and eggs	Gander 6.3–8.2kg (14–18lb) Goose 5.4–7.3kg (12–16lb)	An auto-sexing breed: the colour of the bird reflects its gender. The female is completely grey apart from her stern (backside) and spectacles, with the white increasing around the face with age. The male is mostly white with touches of pale grey on its back, wings and tail. Breeders can sex goslings at hatching, the females having a dark bill and darker down than the males. For the less experienced unequivocal sexing can be done at four weeks when wing feathers will be grey (females) or white (males). Bills, legs and feet of adults are orange. Females have brown eyes, males blue. Lay about 30 eggs per year and make good broodies and parents
Pink-footed	Wildfowl	2.2–2.7kg (4.8–6lb)	Medium-sized goose, pinkish grey with a dark head and neck, pink bill and pink feet and legs. Does not breed in the UK, but large numbers of birds spend the winter here, arriving from their breeding grounds in Spitsbergen, Iceland and Greenland. They eat grain, winter cereals, potatoes and grass
Pomeranian (Pommern Gänse)	Domestic. Medium. Meat	Gander 8.2–10.9kg (18–24lb) Goose 7.3–9.1kg (16–20lb)	With its plump, meaty body, it is bred in large numbers in Germany and Poland to produce smoked goose, salted goose and goose fat. The most common colour in the UK is the grey and white pied/saddleback in white and grey (very similar to the Grey Back), but there are also all grey and all white varieties. Orange-pink bill. Blue eyes and single-lobed abdomen. Hardy breed, and good forager. Lays 30–40 eggs per year. They make good guard birds
Red-breasted	Wildfowl	1–1.5kg (2.2–3.3lb)	Small goose measuring 53–56cm (21–22in) in length. Brightly and strikingly marked. In the wild, breeds in Siberia, but is also common in captive UK wildfowl collections. Lays 3 to 8 white eggs with a green tinge

Goose breeds	Classification and type	Weight of live birds	Information	
Roman**	Domestic. Light. Meat and eggs	Gander 5.4–6.3kg (12–14lb) Goose 4.5–5.4kg (10–12lb)	Pure white, small chubby bird with pale blue eyes and orange-pink bill, legs and webs. Slightly larger than the Czech (Bohemian) goose. There is a crested variety, the tuft smaller than that found in crested ducks. Can be nervous, so interact with them when young to calm them. They are excellent layers, producing 40–65 white eggs per year	
Scanian (Skåne or Skånegås)	Domestic. Heavy. Meat	Gander 7–11kg (15.4–24.7lb) Goose 6.5–8kg (14.3–17.6lb)	The largest Swedish goose. White with brown-grey head, neck, back, thighs and rump feathers. The legs are orange; the bill is orange with a flesh-coloured tip. Of fine meat quality, the skin and meat is white. Easy to fatten, fast growing and robust. Known for having a good temperament. It lays 20–30 grey-white eggs, and is strongly inclined to go broody	
Sebastopol **	Domestic. Light. Ornamental and meat	Gander 5.4–7.3kg (12–16lb). Goose 4.5–6.3kg (10–14lb)	A very distinctive flightless bird with uniquely frizzled/curled feathers, in two varieties: the curled feather and the smooth-breasted (or trailing feather). The feathers on the former are smooth on the head and upper neck, but everywhere else are profusely curled. The smooth-breasted has smooth feathers on the head, neck, breast, belly and paunch; everywhere else the feather is loosely curled and spiralled. The majority are entirely white, but there are grey-flecked, grey, saddleback and buff versions too. Bills and webs are orange-pink. They need access to water to care for their unusual plumage. Not particularly docile. Prone to wing abnormalities. Produce 25–40 eggs per year	
Shetland**	Domestic. Light. Meat and grazing management	Gander 6kg (13.2lb). Goose 5kg (11lb)	Autosexing breed, the males being mainly white and the females grey and white, both with blue eyes. Pink bills, legs and webs. Small and hardy, they are known as a quiet breed. Good foragers, used to rid the grass of parasites, such as the liver fluke, preparing ground for later grazing by sheep. They fly well. They lay around 20 eggs per year, but do not make a reliable broody	

(continued...)

Goose breeds	Classification and type	Weight of live birds	Information
Snow goose	Wildfowl	2.6kg (5.7lb)	Grass-eating medium-sized bird in two colourways: white with black wing feathers, or white-headed with blue-grey body and wings. Breed in Greenland, Arctic North America and Siberia. In the UK they may appear in flocks of white-fronted geese in Scotland and Ireland. A feral flock breeds on the Inner Hebrides
Steinbacher**	Domestic. Light. Meat and eggs	Gander 6–7kg (13–15lb) Goose 5–6kg (11–13lb)	Small attractive goose in blue, grey, lavender, buff, cream and white varieties. The bill is most distinctive: it has a black bean at the tip and black serrations, the remainder being bright orange, as are its legs. Known as a fighting goose, but in fact no more aggressive than other breeds. It lays 5–25 eggs, but the female has an unusually short breeding life, and is unlikely to lay beyond six years of age
Swan goose	Wildfowl	2.8–3.5kg (6.2–7.7lb)	Large goose that breeds in China, Mongolia and Russia. Domesticated flocks now exist in many collections in other parts of the world, including the UK. The Chinese and African geese are descendants of the Swan goose
Taiga Bean	Wildfowl	Gander 2.6–4.0kg (5.7–8.8lb). Goose 2.2–3.4kg (4.8–7.5lb)	The name 'bean' goose comes from the past habit of this goose to graze in bean-field stubbles during the winter. This is the species of bean goose most likely to be seen in the UK. It is typically larger than the related Tundra Bean goose, with similar plumage, a sleeker body and longer neck. They eat grass, cereals, potatoes and other crops

Goose breeds	Classification and type	Weight of live birds	Information
Toulouse**	Domestic. Heavy. Meat and eggs	Gander 11.8–13.6kg (26–30lb). Goose 9.1–10.9kg (20–24lb)	A very large bird; the most commonly known is the grey, but they also come in buff and white varieties. The British and American exhibition birds have a deep keel and large dewlap (the flap of skin hanging beneath the lower jaw); the French and German birds are more upright, with a smaller keel and dewlaps. They have a massive head and a long, broad, deep body with short legs. The legs and webs of the grey are pinkish orange, the bill orange. They are docile, gentle giants, slow growing, and their size can make them challenging to breed. Utility Toulouse are slighter with no dewlap, and are invariably crossbreds. They lay 30–40 white eggs, though some strains produce an astonishing 220–240 eggs per year. Can be susceptible to flystrike (maggots). Seen as a supreme bird for crossing with others to produce meat birds, as pure Toulouse are very fat-laden. They have a shorter lifespan than other geese, averaging ten years, although they can live for twice as long
Tundra Bean	Wildfowl	Gander 1.9–3.3kg (4.2–7.3lb). Goose 2.0–2.8kg (4.4–6lb)	Darker and browner than other grey geese species, with orange legs. Smaller than the related Taiga Bean goose, with a stockier body and shorter neck. It breeds in the Russian tundra, and winters at coastal locations in Europe. They eat grass, cereals, potatoes and other crops
West of England**	Domestic. Heavy meat	Gander 7.3–9.1kg (16–20lb). Goose 6.3–8.2kg (14–18lb)	Autosexing breed, the female showing pied characteristics. The male is mainly white with some grey on the back and rump. The female has a saddleback grey and white pattern, with grey on the head and neck, the grey on the head decreasing with age. Legs and webs are orange-pink. Both have blue eyes. Like the Pilgrim, it can be sexed at hatching: females have grey patches on the beak rather than the plain pale orange of the male. Lays 20–30 white eggs per year
White-fronted	Wildfowl	1.9–2.5kg (4.2–5.5lb)	Grey with large white patch at the front of the head around the beak and bold black bars on the belly. Legs are orange. Siberian birds have pink bills, while Greenland birds have orange bills. Both kinds visit the UK in winter. Does not breed in the UK. They eat grass, clover, grain, winter wheat and potatoes

(continued...)

Duck breeds	Classification and type	Weight of live birds	Information
Abacot Ranger* (Streicher, Hooded Rangers)	Domestic. Light. Eggs and meat	Drake 2.3–2.5kg (5–5.5lb) Duck 2–2.3kg (4.5–5lb)	Lively foraging duck. Both sexes have distinct solid-coloured heads, fawn for the female and black with green lustre for the male. The legs and webs are dark steely grey. Easily sexed from 8 weeks of age, the male has an olive-green bill, the female's is dark slate grey. They are good broodies and layers, producing 180–200 eggs per year. Long-lived and placid
Aylesbury*	Domestic. Heavy exhibition. Eggs: seasonal layer	Drake 4.5–5.4kg (10–12lb) Duck 4.1–5kg (9–11lb)	Bred in Aylesbury as a white table duck, the correct exhibition Aylesbury has a pronounced keel and long pink bill with bright orange legs and webs. The deep keel means that swimming water is required for effective breeding. There are many utility birds without the keel that are Aylesbury/Pekin crosses, most of which will have orange rather than pink bills. Lays 80–100 large white eggs per year (their eggs are often blue in the USA)
Bali	Domestic. Light	Drake 2.3kg (5lb) Duck 1.8kg (4lb)	Originally imported from Malaysia, the breed was re-created in the UK by crossing crested ducks with Indian Runners. Many colours exist, plus white, with orange-yellow bill and orange legs. They look like Indian Runners with a pompom hat (although more rounded over the skull). Because of the crested gene they carry the same problems as the crested duck (see Chapter 9).

Duck breeds	Classification and type	Weight of live birds	Information	
Black East Indian*	Domestic bantam. Eggs: seasonal layer	Drake 0.9kg (2lb) Duck 0.7–0.8kg (1.5–1.75lb)	A bantam duck with the same colour genes as the larger Cayuga. The drakes tend to keep their black-beetle green sheen, while females develop white patches as they age. Bills are black, and legs and webs too. Excellent fliers, so will need their wings clipping. Egg colour grey, which fades to white as the season progresses. They lay 40–100 eggs per year	
Call (Decoy)	Domestic bantam. Eggs: seasonal layer	Drake 0.6–0.7kg (1.25–1.5lb) Duck 0.5–0.6kg (1–1.25lb)	The smallest domestic duck, used for centuries to attract wild ducks into traps; their characteristic highly vocal trait was an important lure. Rounded head, short bill and chubby cheeks to match their compact bodies. Seventeen colour variations, plus crested types. Excellent fliers so will require clipping. Although small, they have tasty meat. Lay 20–60 white eggs per year, and make decent broodies. Their constant calling can attract predators and be a challenge for neighbours	
Campbell*	Domestic. Light. Eggs	Drake 2.3–2.5kg (5–5.5lb) Duck 2–2.3kg (4.5–5lb)	Campbells rarely fly, and come in khaki, white and dark colours. Agile, fertile and prolific, there are many variants beyond the white and dark including the Welsh Harlequin, the Abacot Ranger (a cross back to the Runner) and the Whaylesbury hybrid (Harlequin and Aylesbury). The Khaki Campbell is the most successful egg-laying utility breed, producing an astonishing 300–350 white eggs per year. They are of no use as broodies. Good quality lean meat	
Carolina (Wood)	Wildfowl	454–862g (1–2lb)	North American crested duck. Lays 7–15 eggs, although nests may have up to 40 eggs from multiple females. Agile tree climber that nests in hollow trees. Very attractive birds with a range of colour variations	

(continued...)

Duck breeds	Classification and type	Weight of live birds	Information
Cayuga*	Domestic. Heavy exhibition. Eggs: seasonal layer, meat	Drake >3.5kg (8lb) Duck >3kg (7lb)	North American origins, a black bird with beetle-green sheen, its colouring is very similar to the smaller Black East Indian. Bill and legs are black, with some orange as drakes age. Ducks in particular turn increasingly white with age. Eggs can be very dark early on, but the shell is white under the sooty pigment, which washes off. Lays 80–100 eggs per year. Swimming water helps them to stay healthy; the oil that produces the sheen will dirty their water
Common Scoter	Wildfowl	650g–1.3kg	A sea duck, the male is totally black with a yellow dab on the bill, and the female lighter with a pale face. Seen offshore or in long lines flying along the coast. They eat molluscs
Crested*	Domestic. Light exhibition. Eggs: seasonal layer	Drake 3.2kg (7lb) Duck 2.7kg (6lb)	These birds have a significant pompom (crest) of feather on their heads. Comes in all white and other colour variations. Be aware that there are serious welfare issues with all crested breeds (*see* Chapter 9)
Eider	Wildfowl	1.2–2.8kg (2.6–6lb)	The UK's heaviest wild duck and the fastest flying. A sea duck rarely found away from the coast. Eats shellfish, particularly mussels

Duck breeds	Classification and type	Weight of live birds	Information
Gadwall	Wildfowl	650–900g (23–32oz)	Grey dabbling duck with a black rear end, smaller than the mallard. The grey colour is barring and speckling when seen up close. Visits gravel pits, lakes, reservoirs and coastal wetlands and estuaries in winter
Garganey	Wildfowl	250–450g (8.8–15.8oz)	Rare, secretive breeding duck smaller than a mallard and slightly larger than a teal. The male is most easily recognized with a broad white stripe over the eye. Eats plants and insects
Goldeneye	Wildfowl	650g–1.2kg (23oz–2.6lb)	Medium-sized diving duck. Males are black and white with a greenish-black head and a circular white patch in front of the yellow eye. Females are smaller, and mottled grey with a chocolate-brown head. They eat mussels, insect larvae, small fish and plants
Goosander (Common Merganser)	Wildfowl	Drake 1.3–2.1kg (2.8–4.6lb). Duck 900g–1.7kg (32oz–3.7lb)	Diving ducks, and a member of the sawbill family because of their long, serrated bill, used for catching fish, in particular trout and salmon. A largely freshwater bird. Form into flocks of several thousand in some parts of Europe

(continued...)

Duck breeds	Classification and type	Weight of live birds	Information
Hookbill*	Domestic. Light exhibition. Eggs (seasonal layer) and meat	Drake 2–2.5kg (4.5–5lb) Duck 1.6–2kg (3.5–4.5lb)	Distinctive, long curved bill. Three colour variations: the dusky mallard, the white bibbed and the white. Flies well and goes broody. Can lay over 100 eggs per year. Egg colour blue
Indian Runner	Domestic. Light. Eggs	Drake 1.6–2.3kg (3.5–5lb). Duck 1.4–2kg (3–4.5lb)	Unique upright stance (not unlike a wine bottle), long and slender. Comes in fourteen colour variations in the UK. Crossed with other breeds of domestic waterfowl, the runner has produced nearly all the light duck breeds including the Khaki Campbell and Orpington, all of which have a lower carriage than the standing tall runner. Used on organic farms and vineyards for pest control, and also popular as pets. Not a flying breed. Very moderate appetite. Lay 200–250 eggs per year. Egg colour blue-green to white, although early eggs from black runners can be very dark grey
Long-tailed	Wildfowl	520–950g (18–33.5oz)	Small sea duck with a round head. In winter the male is mainly white, with brownish-black markings and greatly elongated tail feathers. Females are browner. They do not breed in the UK but are a winter visitor. Eats mussels, cockles, clams, crabs and small fish
Magpie*	Domestic. Light. Eggs and meat	Drake 2.5–3.2kg (5.5–7lb). Duck 2–2.75kg (4.5–6lb)	Welsh breed, with black and white, blue and white, and dun and white varieties with orange legs. The head has a coloured cap above the eyes, to the back, tail and shoulders in a heart shape across the back, with the remainder being white. The magpie patterning was developed so that no dark stubs would show on its breast when plucked. Bill colour in the first year should be yellow, and in the duck turns to a shade of cucumber green in the second year, while the drake may develop spots of green in later years. Can fly fairly low to the ground. Lays up to 200 eggs per year, egg colour from white to pale blue-green

Duck breeds	Classification and type	Weight of live birds	Information	
Mallard	Wildfowl	750g–1.5kg (26oz–3.3lb)	The most common wild duck. Large with a long body, and long broad bill. The male has a dark green head, a yellow bill, is mainly purple-brown on the breast and grey on the body. The female is mainly brown with an orange bill. Mallards breed in all parts of the UK in summer and winter, wherever there are suitable wetland habitats. They eat seeds, acorns and berries, plants, insects and shellfish	
Mandarin	Wildfowl	430–690g (15–24oz)	The male has beautiful ornate plumage with distinctive long orange feathers on the side of the face, orange sails on the back, and pale orange flanks. The female is much duller with a grey head and white stripe behind the eye, brown back and mottled flanks. Introduced to the UK from China, and became established following escapes from captivity. Eats insects, vegetation and seeds. Found on lakes with plenty of overhanging trees and bushes	
Muscovy	Domestic. Heavy. Eggs and meat	Drake 4.5–6.3kg (10–14lb). Duck 2.3–3.2kg (5–7lb)	The domesticated Muscovy is the only breed of duck not descended from the wild Mallard; they are not true ducks, but are perching waterfowl. They forage grass like geese, and eat large numbers of insects. The females are half the size of the males. They come in many colour variations, with distinctive, red, bare facial patches, and wart-like caruncles that develop as they age; these are larger in drakes. Prized as a broody that will effectively rear their young. Fly very well for a heavy bird, and will roost in trees and on roofs, holding their position with strong claws; wing clipping may be needed to contain them. Noted as a quiet breed. Lay 100–180 eggs per year; egg colour pale olive to white. Incubation 35 days rather than the usual 28 days for ducks. Their meat is highly prized, and known as Barbary duck	
Orpington**	Domestic. Light. Eggs and meat	Drakes 2.2–3.4kg (5–7.5lb). Ducks 2.2–3.2kg (5–7lb) (top of the size range for a light breed)	Developed by William Cook of Kent (who also bred Orpington chickens) in the late nineteenth century, believed to be the result of cross-breeding Indian Runners to Aylesburys, Rouens and Cayugas. There are breed standards for Buff and Blue varieties, but Black, White and Chocolate birds also exist. Hardy and active, producing a good-sized table bird, and a good egg layer with an average of 150 eggs per year. Egg colour white	

(continued...)

Duck breeds	Classification and type	Weight of live birds	Information
Pekin**	Domestic. Heavy. Meat	Drakes 4.1kg (9lb) Ducks 3.6kg (8lb)	Originates from China, although the popular bird with upright penguin stance is known as the German Pekin (the American Pekin is nothing like as upright). Thick neck and white plumage with a yellow sheen, yellow bill and orange legs. An exhibition breed and important base stock for commercial breeding for table birds. Lays around 200 eggs per year. Egg colour white
Pintail	Wildfowl	0.55–2.2kg (1.2–4.8lb)	Slightly bigger than a mallard, a long-necked, small-headed duck with a curved back, pointed wings and a tapering tail
Pochard	Wildfowl	930g (33oz)	In winter and spring, male pochards have a bright reddish-brown head, a black breast and tail, and a pale grey body. Females are brown with a greyish body and pale cheeks. They eat plants and seeds, snails, small fish and insects
Red-breasted Merganser	Wildfowl	900g–1.35kg (32oz–3lb)	Diving ducks of the sawbill family, with a long, serrated bill, used for catching fish such as salmon and trout. At home on both fresh and saltwater; most commonly seen around the UK's coastline in winter. They form flocks of several hundred in the autumn

Duck breeds	Classification and type	Weight of live birds	Information
Rouen*	Domestic. Heavy exhibition. Eggs (seasonal layer)	Drake 4.5–5.4kg (10–12lb). Duck 4.1–5kg (9–11lb)	Originates from northern France and used as the basis for breeding many of the duck breeds developed in the twentieth century. Large, horizontal bird with deep keel parallel to the ground. Too heavy to fly. Bill in the drake is a bright green-yellow, the duck is orange with a black saddle and bean, its legs orange. There is also a blue form, but as with other blue ducks, this does not breed true. Lays between 100 and 150 eggs per year. Slow growing, taking two years to reach its full weight of 5kg. Egg colour white	
Rouen Clair (Duclair)	Domestic. Heavy. Meat	Drake 3.4–4.1kg (7.5–9lb). Duck 2.9–3.4kg (6.5–7.5lb)	Traditional mallard colouring, a little more upright than the Rouen, and a long bird. Tendency to get fat, so needs room for foraging and exercising, and avoid overfeeding. Docile, and does not fly well. Lays more eggs than the Rouen, at 150–200 per year. Egg colour pale blue-green to white	
Ruddy Duck	Wildfowl	350–800g (12–28oz)	Small, stout, freshwater diving birds. The male ruddy duck has a bright chestnut body, black crown, white cheeks and blue bill. They swim with the tail cocked up, and submerge gradually without diving. They rarely leave the water, being very ungainly on land. Categorized as an invasive species, and a risk to the native White-headed duck	

(continued...)

Duck breeds	Classification and type	Weight of live birds	Information
Saxony	Domestic. Heavy. Meat and eggs	Drake 3.6kg (8lb) Duck 3.2kg (7lb)	Bred in the 1930s from German Pekin, Blue Pomeranian and Rouens. Large, active bird. Bill yellow, legs orange. Egg colour white. Lays 80–100 eggs per year. Not a broody breed. Females are noisy
Scaup	Wildfowl	0.8–1.3kg (1.7–2.9lb)	Diving ducks. Males have black heads, shoulder and breast, white flanks, grey back and a black tail. Females are brown, with characteristic white patches around the base of the bill. They eat shellfish, crustacea and small insects. A handful breed in the UK every year, making them our rarest breeding duck
Shelduck	Wildfowl	0.85–1.4kg (1.9–3lb)	Large, colourful duck. Both sexes have a dark green head and neck, a chestnut belly stripe and a red bill. They eat invertebrates, small shellfish and aquatic snails. Found mainly in coastal areas, though they can also be found around inland waters such as reservoirs and gravel workings
Shetland**	Domestic. Light. Eggs	Drake 2kg (4.4lb). Duck 1.8kg (4lb)	Critically endangered, native to the Shetland Islands in Scotland. Tough, active birds. Black with green/ blue sheen with white breast and occasional spotting. Drakes have yellowish bills and orange on their legs, the females have grey or black bills and legs. Ducklings are dark grey with beige or cream bib and dark blue bills that lighten with age. As with Cayuga, they fade in the females with age and revert to white. Prolific layers, around 150 eggs per year, egg colour varies from white to grey

Duck breeds	Classification and type	Weight of live birds	Information	
Shoveler	Wildfowl	400g–1kg (14oz–2.2lb)	Surface-feeding ducks with huge spatulate bills. Males have a dark green head, with a white breast and chestnut flanks. Females are mottled brown. In the UK they breed in southern and eastern England. They eat small insects and plant matter sifted from the water	
Silver Appleyard*	Domestic. Heavy. Eggs and meat	Drake 3.6–4.1kg (8–9lb). Duck 3.2–3.6kg (7–8lb)	Produced by Reginald Appleyard from selective cross breeding of Rouen, Pekin and Aylesbury. Large, long-bodied bird. Bill yellow for male and female, legs light orange, blue wing bar. Body colour is a lighter version of the basic mallard pattern. Bred as a dual-purpose bird: prolific egg layer (around 160 up to 250 per year), and a quick-growing table bird with a deep meaty breast. Egg colour white	
Silver Appleyard Miniature	Domestic. Bantam. Eggs (seasonal layer)	Drake 1.4kg (3lb) Duck 1.1kg (2.5lb)	Miniature version of the Silver Appleyard (a third of the size), developed in the 1980s. Small, slender, not as noisy as the call duck. Long yellow bill, orange legs. Egg colour white (blue tinge); lays 60–160 eggs per year	
Silver Bantam**	Domestic. Bantam	Drake 0.9kg (2lb) Duck 0.8kg (1.75lb)	The drake follows the mallard colour for the head and neck, the breast is claret with white edges. Under-body is cream, the back black with white edges. Olive-green bill and orange legs. The duck has a brown head and neck, the body is a creamy white with a central stripe on the feathers of light brown. Dark slate bill tinged with green, dark orange legs. Lays 60 to 160 eggs a year, sits well, is an attentive mother	

(continued...)

Duck breeds	Classification and type	Weight of live birds	Information
Smew	Wildfowl	500–800g (17.6–28oz)	Small diving duck with a delicate bill. The male is white with a black mask and a black back, the female is grey with a reddish-brown head and white cheek. A winter visitor in small numbers from Scandinavia and Russia
Stanbridge White**	Domestic. Light. Eggs.	Drake 2.5–3.2kg (5.5–7lb) Duck 2–2.7kg (4.5–6lb)	Believed to be extinct until 2007. Pure white with a yellow/orange bill and yellow legs. More upright than the Aylesbury, same shape and size as the Magpie (from which they may have originated), but with a chunkier breast. Excellent layers, producing 240–250 eggs per year, lay from February into October. Medium sized, the breed can produce good meat birds. Egg colour: blue-green hue at the beginning of the season, fades to grey pearl
Swedish Blue	Domestic. Heavy exhibition. Lays well	Drake 3.6kg (8lb) Duck 3.2kg (7lb)	Dark blue bill, legs browny orange, black head, white bib and blue body. Offspring may be silver, black or brown; only 50 per cent from two Blue Swedish parents will be blue. Calm birds, good foragers. Females known to be noisy. Egg colour pale blue-green. Lays 100–150 eggs per year
Teal	Wildfowl	240–360g (8.5–12.7oz)	Small dabbling duck with many varieties. The common Eurasian Teal male has a chestnut-coloured head with broad, green eye patches, a spotted chest, grey flanks and a black-edged yellow tail. The female is mottled brown. In winter found in wetlands in the south and west of the UK. They eat seeds and small invertebrates

Duck breeds	Classification and type	Weight of live birds	Information	
Tufted duck	Wildfowl	450g–1kg (15.9–2.2lb)	Medium-sized diving duck, smaller than a mallard. It is black on the head, neck, breast and back, white on the sides, with a small crest and yellow eye. Breeds in the UK across lowland areas of England, Scotland and Ireland. They eat molluscs, insects and plants	
Velvet Scoter	Wildfowl	1.1–2kg (2.4–4.4lb)	Black sea duck with a long bill, thick neck and pointed tail, with a white patch on the rear of the wing. A winter visitor to the east coast, especially in Scotland, Norfolk and north-east England. They eat shellfish, crabs, sea urchins, fish and insect larvae. None are kept in captivity in the UK	
Welsh Harlequin**	Domestic. Light. Eggs and meat	Drake 2.3–2.5kg (5–5.5lb). Duck 2–2.3kg (4.5–5lb)	Dual-purpose breed known for its outstanding laying ability. The drake's head is greenish black, the duck has a creamy white head with brown stippling. Good foragers, they produce lean white meat and lay 240–330 eggs per year. Egg colour white	
Wigeon	Wildfowl	500–900g (17.6–31.8oz)	Medium-sized with a round head and small bill. The head and neck of the male are chestnut, with a yellow forehead, pink breast and grey body. They breed in Scotland and northern England. They eat aquatic plants, grasses and roots	

CHAPTER SEVEN

Buying Your First Ducks and Geese

Looking at lists of duck breeds, you might be persuaded to make a choice based on the descriptive name of their colour alone – though butterscotch, lavender, black ripple and chocolate sound more like pudding recipes than bird colours. Nevertheless, the concept of having birds that approximate to the best aspects of cake or ice-cream flavours may seem comic, but I've heard of worse plans. Choosing rare or exotic breeds that require expert care when your only bird experience is a childhood budgie may be something to aspire to, rather than to start off with.

What's important is that you take into consideration space (*see* Chapter 4), your reason for keeping waterfowl (*see* Chapter 2), the commitment you can give them, their availability, and how to choose healthy stock that will do well for you. Do also consider seasonality: asking for point-of-lay birds at the beginning of the year is unlikely to yield success. I've frequently been asked to sell ducklings as Easter presents, and the answer is always a firm 'No', as I won't have birds available at that point, and more importantly, it is highly likely that they will be swiftly dumped once the novelty has worn off.

WHAT WILL SUIT YOU BEST?

In Chapter 2 I talked about choosing a type or breed of livestock as being very much a personal thing. The more you are drawn to

OPPOSITE:
Welsh Harlequin duckling.

RIGHT:
Exmoor, October 1985.
Documentary photo: James Ravilious/Beaford Archive

White Runner ducks at a show and sale.

certain birds, the more effective you are likely to be as their keeper. You may be charmed by the upright stance of the Indian Runner and appreciate their good egg-laying ability, or aspire to excellent meat birds and prefer the Rouen to the Aylesbury in looks, or be thoroughly impressed by the grandeur and pomp of Toulouse geese or the curly feathering of the Sebastopol. If you are unfamiliar with the possibilities, read through the guide to the breeds in Chapter 6, and make an assessment as to what will suit you best.

Birds that are Fit for Purpose

If you want to be self-sufficient in duck eggs, avoid the small and not hugely productive Call duck, and choose Welsh Harlequins or Campbells, Pekins or Stanbridge Whites, Magpies or Indian Runners instead. If you want meat and eggs, you have a great range of dual-purpose ducks to choose from, including Swedish Blue, Silver Appleyard, Saxony, Rouen Clair, Rouen, Pekin, Muscovy, Cayuga and Aylesbury.

For geese, there is less of a distinction in egg-laying abilities, with most laying around thirty eggs a year, although the prolific Chinese goose can achieve the heady heights of up to eighty if you're lucky. The more significant anserine (goosey) difference is in weight and stature, from the same prolific but modestly sized Chinese at around 4.5kg to the mighty Embden weighing more than three times that and standing a metre tall.

If you are less interested in the production side of the birds but want something more pet-like, very young birds will enable you to handle them and get them fully acclimatized to being happy around humans. Ducks that are considered particularly tame include Pekin, Aylesbury, Khaki Campbell, Call, Muscovy and others, but this characteristic is determined more by gentle handling and time spent with them than the breed.

The same is equally true of geese – hand rearing young birds is essential if you are hoping for pets. Toulouse are considered calm, as are Pilgrim, Embden, Buff, Sebastopol and Roman (and the list could be extended to most, if not all our domestic breeds). However, be aware that in the breeding season they are far less likely to be peaceful and docile, and even if they are calm around their own human, this does not mean they will behave in the same

Birds or Hatching Eggs?

Should you choose birds or hatching eggs to start your flock? The following table sets out the advantages and disadvantages of both routes.

Pros and cons	Buying birds	Buying hatching eggs
Acclimatization to their human	If a few days old, there should be no problems in raising friendly birds	You will be handling the chicks from day one
Health status	There can be a risk of bringing in disease or pests – ensure sound quarantine practices (*see* Chapter 10)	Generally understood to be the healthiest option for not introducing disease to your flock, although not infallible
Gender mix	You have the option of buying sexed birds, although this is difficult for many breeds until they are at least adolescent, unless the seller is reliably skilled at vent sexing or you are buying autosexing breeds	You have no control over the gender of your birds – what will hatch will hatch, with a probability of 50:50 that each egg will produce a male or female. If you end up with excess males you will need a plan for them: sell/keep/rear for meat
Knowledge	It is easier to rear birds than to hatch eggs	There is enjoyment of course, but also skill involved in hatching your own birds
Natural incubation	It is not advisable to introduce young birds to adults – they are very much at risk, so keep them separated	If you have a broody bird, whether duck, goose or chicken, she can hatch eggs for you
Artificial incubation and provision of heat	No incubator required, but young birds will need to be kept under heat in a brooder (*see* Chapter 12)	You will need an incubator, brooder and heat lamp (*see* Chapter 12)
Certainty of numbers	If you buy healthy stock you should have some degree of certainty	There are many reasons why eggs do not hatch; it is very rare to get a 100 per cent hatch rate
Availability	If you are fastidious about the breeds you want, you may find that only adult birds are available	Fertile eggs are only available in season, and may have to be posted if located at a distance; rough handling by the courier has a detrimental effect on hatchability
Expense	The less expensive route even though the initial purchase price is higher	The more expensive route once the costs of incubator, heat lamp and electricity are included

way with others. There are plenty of diverse comments made about geese, ranging from 'aggressive' to 'make great pets' for the same breed – but only the fact that you rear them yourself can really make a difference for much of the year. That being said, although geese will warn you off with hissing and a determined approach, a goose will rarely attack unless genuinely provoked.

If conservation is of interest, you have plenty of choice, because many of our domestic breeds of ducks and geese are acknowledged priority breeds, as identified by the Rare Breeds Survival Trust (*see* asterisked breeds in Chapter 6).

SOURCING QUALITY BIRDS AND EGGS

The lure of the bird auction can be strong, and if you have never attended a livestock auction it is easy to get carried away, either buying too many birds or bidding more than you should, although there are frequently bargains to be had. It's an enjoyable activity, but be aware that although the birds you've set your heart on may look in good health, they may be in cages alongside unwanted birds, and there is the chance that they will bring home parasites and diseases you really don't want to introduce. In addition, there is no follow-up support or guidance from the breeder or the auctioneer.

The best option by far is to buy young waterfowl or fertile eggs for hatching direct from the breeder, who can offer advice and give you all the information you need to know about their birds, including a tour of their pens and facilities. There is also a good chance that you will learn other important things about waterfowl from an experienced keeper, as most people love to talk about their birds. There are breeders' directories at the back of smallholding and poultry magazines, online information on the British Waterfowl Association website and other forums, and local smallholding and poultry groups can signpost you to reputable breeders. It is generally considered that the best way to minimize the risk of bringing in unwanted disease is to avoid purchasing adult waterfowl.

Wherever you source your birds, commit to isolating all new birds from any you already have for a month, so have a separate space for them organized before you bring them home.

Hatching eggs are available from various online sales sites, and many are from reputable breeders – although tales of emerging hatchlings being of an entirely different breed from that ordered are legion. Be aware that boxes roughly handled by a courier or postal system does impact on a successful hatching of the eggs, but if the eggs you want are at a considerable distance from you, it is worth taking the risk;

Call ducks at auction.

I have hatched many birds successfully (and of the breed ordered) from posted eggs that have travelled from the other end of the country.

It is not possible to buy eggs that are a few days away from hatching. Removing them from the incubator so they can be posted elsewhere, or transported to another incubator miles away, will kill the foetus. No breeder would agree to sell late-stage incubated eggs, although they will be asked to do this with surprising frequency.

Many schools incubate duck eggs each year, which can prove an excellent learning opportunity for children, covering the topics of biology, reproduction, food production, art, maths, caring for young animals and more. However, before embarking on a programme of incubation, part of a school's responsibility is ensuring that any resulting birds will have a good home once they are ready to leave the classroom.

BRINGING YOUR NEW BIRDS HOME

Bird Transport Containers

I still remember one couple coming to collect a clutch of two-week old ducklings from the farm one summer. I asked them to bring a carrier or box with them, as I had run out. As they got out

ABOVE: *Poultry crates.*

LEFT: *Transporting geese safely.*

of their car, one of them was holding a shallow biscuit tin, which I assumed contained their lunchtime sandwiches, or perhaps a few sausage rolls and a handful of boiled eggs. But no, they had intended to take their ducklings home in a biscuit tin. It was about a third of the height of the birds, had no ventilation whatsoever, and being made of metal was already hot to the touch from travelling on the car's parcel shelf; I can't imagine a more inappropriate and deadly container.

Cat carriers make excellent bird transport containers: they have good ventilation, are easily hosed out, have a carry handle, and keep the birds secure. Large cardboard boxes work well for big birds such as adult geese, and it's easy to cut ventilation holes; just be sure to do this before you put the bird in. Be sure to properly secure the top of the cardboard box with string or gaffer tape, as simply folding down the flaps will certainly result in an emerging goose, probably trying to peck your ear as you drive down the motorway.

The bottom of the box will get wet – something to be aware of if your car has de luxe cream leather seats – so put plastic sheeting underneath to avoid staining. You can buy purpose-made cardboard animal carriers, but any sturdy cardboard box of the right size will do well for one-off journeys. Dog crates are also suitable as long as the bars of the crate don't allow the birds to escape or to get bits of their anatomy trapped. If you are moving young birds regularly you may want to choose a poultry transport crate – although these are not suitable for tall birds as they are quite shallow, at around 30cm (12in) high.

Whatever container you use, it should be large enough for the birds to stand up and turn round. Make sure ventilation holes are adequate, but not so large that a goose can poke its head through and into all sorts of trouble: geese will test and nibble at everything within their reach. Line the base of the box with newspaper, and then a layer of wood shavings or straw. Don't put feed or water in the box with the birds, but bring some for the journey just in case of delays. Put the container out of direct sunlight, and avoid transporting birds when the weather is very hot or very cold. Take the birds straight home without any detours, and if you have to stop, make sure it's brief and that the vehicle remains ventilated and is parked in the shade.

ON ARRIVAL

The stress of transporting your new birds a long distance is as much trauma as you might want to cause them. When you arrive, don't wrestle them out of their crate, simply put the opened transport box in their new area so they can emerge at will, into a pen with plenty of bedding, water and feed, and leave them be. If it's the evening, put them straight into their huts and let them sleep. Give them a day or two to get used to you and their new surroundings before doing any further husbandry tasks, such as wing clipping. However, if it was a short, easy journey, you might feel happy to get on with wing clipping immediately for flighty breeds (*see* section on wing clipping below).

Introducing New Birds to your Environment and other Birds

It is advisable not to put your new birds in with an existing flock on their arrival, as this would present a significant risk to the newcomers. Your flock will have developed immunity to low levels of pest and pestilence that are peculiar to your territory and set-up. Quarantine your new birds for a month by putting them in clean, disinfected housing and runs on fresh grass; this way they can start to build their own immunity. (For information on quarantine, *see* Chapter 10.)

Some birds, particularly geese, can be very territorial, and even if you put runs close together post-quarantine, or place a cage run into an existing pen for a week or two, giving new and old birds plenty of time to get to know each other through fencing, introducing them into the same area may not work. However, it's always worth a try, particularly if you put both lots of birds into a new space where neither has prior territorial ownership – but be on hand to

How to carry a goose.

separate them if necessary. Introductions are more likely to succeed if you introduce a good-sized group, rather than one or two birds that the old hands can intimidate more easily.

Bringing in a new gander or drake to a group of females is usually an easy call, but adding a new goose to a breeding group is not so straightforward. If you need to do this, remove the gander and put the new female(s) into a cage run inside the existing females' pen for a week, before letting them interact fully. If this works, you can reintroduce the gander to his enlarged harem.

I have had equal rates of success and failure regarding amalgamating flocks, so ensure that you have the space and facilities to handle failure.

Cats, Dogs and other Livestock

Young birds are potentially prey for everything from rats to buzzards (*see* Chapter 10 on keeping your flock secure), but once beyond the fluffy gosling and duckling stage, birds are normally able to fend for themselves as far as cats and smaller predators are concerned. However, if you are bringing home young birds, do make sure you have cat, dog and other predator-proof runs ready for them. For older, more free-

Muscovy duck and duckling with a well-behaved dog.

Dogs and geese.

ranging birds, cats might be curious, but on the whole are unlikely to cause harm.

Dogs, however, can be another matter. Our dogs will lie quietly as I move ducks from their hut and across the farmyard to their daytime run, because I'm there to manage the situation and the dogs know they are not allowed to chase or pester the birds. Even so, I wouldn't leave the dogs with the birds unsupervised for a second. Dogs are predators, birds are prey, and it's not fair to either to leave them alone together – and this is as true for small dogs as it is for large ones. Geese being far more protective are likely to keep dogs at bay, but again, I'd never leave them together. Geese can be quite persistent in hissing and defending themselves, but a dog can only take so much pecking before it decides to prove its superior strength, and ultimately the dog will always win.

Waterfowl and sheep or goats seem to coexist fairly happily as far as temperament is concerned. However, there are potential health risks in grazing alongside each other's faeces, and if mammals or birds have had chemical treatments such as wormers, this is something you can't police unless you keep the species separate. The natural curiosity of goats may also lead to birds being pestered, in which case you would need to separate them anyway. On balance, unless you have spacious shared grassed areas that allow plenty of clean grazing for all concerned, you are probably better off keeping species separately. I'd never keep waterfowl in the same paddock as pigs: the chance of the birds being eaten is far too high, and pigs will quickly churn up any available grazing.

Putting to Bed at Night

For a few nights it can be a little time-consuming putting new birds to bed, as everything is strange and a little confusing for them. However, I've always found geese and ducks far easier to put away at night than chickens or turkeys; it's a simple case of ushering them into their hut (a wide door makes this much easier). Waterfowl all seem keen to go into warm, dry, nicely bedded huts as the light starts to wane, and they will soon be asking to go to bed. The time we do this changes with the alteration in daylight hours. In the summer it might be nearly 9pm before bird bedtime, but in the middle of winter this won't be much later than 3.30pm.

Catching and Handling Ducks and Geese

DUCKS

Poor catching and handling can easily lead to ducks becoming lame; carry them individually, and never catch a duck by the leg as there is a high risk of hip dislocation.

Place a hand on each side of the body over the wings, and lift the bird. Alternatively, catch and gently lift by the base of the neck for a minimal time before transferring the bird to your hand and arm. Holding briefly by the neck minimizes flapping and reduces injury. Take care not to apply excessive pressure when lifting ducks by the base of the neck.

Once you have hold of the bird, slide one hand under the body and firmly clench the legs between your outstretched fingers (positioning one or two fingers between the legs), and support the bird's breast on the palm of the same hand. The wings can then be controlled by your opposite hand or by holding the bird against your body, under your arm. Ducks have sharp claws so it is recommended to wear gloves when handling.

GEESE

To handle geese safely, it is important to have control of the head to avoid being bitten. Geese may be caught from behind by the neck, taking care not to apply excessive pressure. Most geese will sit down once caught in this way, preventing the wings from hitting the handler's legs. Once you have hold of the neck, slide your other hand under the body and firmly clench the legs between your outstretched fingers (positioning one or two fingers between the legs), and support the bird's breast on the forearm of the same hand. The head and neck can then be tucked under your armpit of the supporting arm and the bird lifted against your body. The wings can then be controlled by your free hand.

Picking up a large duck.

Holding a duck's legs in the proper position for carrying.

Catching a goose gently.

Holding and restraining a goose humanely.

Wing clipping.

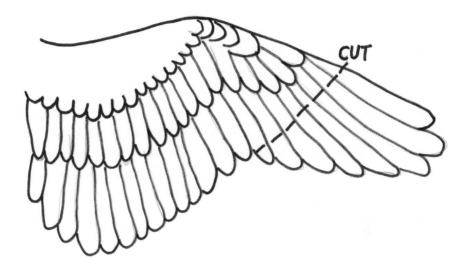

FURTHER HUSBANDRY TASKS

Wing Clipping

Most domesticated duck breeds don't fly, with the exception of the Mallard, Call, East Indies and Muscovy. As for geese, most can fly to a greater or lesser extent, although I've only known our Pilgrims fly any distance after sitting and hatching a brood, having lost a significant amount of weight in the process. If your birds are non-flyers, there is no need to wing clip.

Wing clipping is the removal of the tips of the primary wing feather on one wing, so as to unbalance the bird so that it can't fly. It's a painless process, much like trimming your fingernails or having your hair cut. If you are doing a number of birds, decide to do either all the left wings or all the right, or you may pick up a bird already clipped and do both sides inadvertently. If you do both wings by mistake, the goose might not fly as well as it could before, but it can still fly. Use sharp scissors to avoid hacking at the feathers. Trimming lasts only as long as the next moult, so you will have to do this annually when the feathers have regrown.

Be careful to trim only the primary flight feathers, and avoid the wing tip, which is flesh and bone: it's easy to tell feather from flesh if you feel for the wing tip before you start to cut. If you wish, you can leave three or four of the large flight feathers at the tip of the wing so it looks more normal when folded against the body. For ducks, wait until they are fully feathered at fifteen weeks old, and geese at seventeen weeks before you clip one of their wings.

Sexing Your New Birds

Most ducks apart from the Muscovy can be sexed by voice by the time they are fully feathered at five or more weeks: the female will give the ubiquitous 'quack' expected of ducks, while the male has a raspier sound, which tends to be more insistent. To be sure that the sound you are hearing is from a specific bird, pick it up, as handling often encourages it to quack or rasp. The Muscovy male is older by the time you can voice sex, when he starts to hiss at around sixteen weeks. At sixteen to twenty weeks drakes will develop distinctive curled tail feathers – apart from the Muscovy who doesn't sport curled

Autosexing Pilgrim goslings: three males on the left and three females on the right, clearly showing the yellow/pale grey of the male, and the darker olive-grey colour of the female and their temporarily dark bills.

tail feathers. Note that unscrupulous sellers can catch the inexperienced duck keeper unawares by removing the curled tail feathers: knowing how to voice sex your ducks will keep you from buying a group of drakes masquerading as layers.

Many duck breeds have significant colour and pattern differences between the male and female, making them easy to tell apart, just like their ancestor the wild mallard, where the male is showy with an emerald head and the female is less vibrantly coloured, speckled brown. However, this is no help for birds such as the Pekin or Aylesbury.

The colour of the bill can be helpful in indicating gender. In some breeds (Rouen, Appleyard, Welsh Harlequin) the duck has a dark orange or brown bill, while the male has a green bill – but this is not a general rule, as in some other breeds the female has a green bill as well as the male.

The colour of the day-old baby fluff in some breeds of gosling (and the colouring of their feathering as adults) can be used to distinguish the gender in autosexing breeds, where the colour of the bird indicates the gender. Autosexing birds include Pilgrim, Shetland, West of England and

Muscovy male and female mating.

Normandy geese. Otherwise, telling the sex of geese can be challenging until they are several months of age when their behaviour starts to give clues that indicate gender.

Unlike toads and spiders where the females are predominant, drakes and ganders are larger than their female counterparts, sometimes significantly so, as in the case of the Muscovy, where the drake can be twice the size of the female, and has larger fleshy caruncles around the face. For all breeds, the males tend to have larger feet than the females. Adult ganders have longer, thicker necks and behave more aggressively, but plenty of females show protective behaviour, so don't go by that alone. The knobs on African and Chinese geese will develop earlier and become more prominent in the male.

Vent sexing (checking for the penis manually by gently opening the vent of young birds, preferably day-olds) is a real skill and tends to be restricted to those raising commercial birds for fear of inexperienced hands unwittingly damaging a bird.

CHAPTER EIGHT

Feeding and Nutrition

THE WATERFOWL GIZZARD

Geese and ducks definitely nibble, but of course they don't actually have teeth. Bills are their nibblers and shovellers, and the part that does the equivalent of chewing is the gizzard, a tough muscle that receives the food from the oesophagus and passes it into the small intestine, grinding up the food so that the nutrition can be extracted. If you ever eviscerate a bird, take time to explore the gizzard. If you slice it in half (which you'll need to do in order to clean it if you intend to include it in the giblet stock for the best gravy) you'll find it full of grit and feed in various stages of digestion. The inner layer that encases the food and grit is yellow, and can be easily peeled off and discarded.

LABELLING AND STORAGE

Poultry grower and layer pellets look identical, so when you store feed, make sure you label the containers with the food type as appropriate, to avoid misfeeding your birds. It is also essential to use something vermin proof for storage. Metal dustbins work well (although I once found a pair of rotund woodmice in a metal, lidded feed bin and still have no idea how they got in there), as do out-of-use chest freezers, though be sure to disable and remove any locks and electric cables to make freezers child safe. Don't store feed in their plastic or paper sacks on pallets, or anywhere else where rodents can gnaw through the sacks and contaminate and spill the feed. You really do need to transfer

OPPOSITE: *Geese eating weeds from the vegetable garden.*

RIGHT: *Storm cloud with geese, Swimbridge, 1986. James Ravilious/ Beaford Arts*

Mixed poultry corn.

Poultry grower pellets.

Poultry layer pellets.

the sacks or their contents into something that cannot be penetrated by rat and rain.

Buying feed in bulk to save money and not having adequate storage for larger quantities is a false economy. Be aware that bird feed, just like our own, has a 'use-by' date, so be sure not to top up feed bins by putting new food on top of old. The industry standard use-by date is three months from the date of manufacture and bagging; however the vitamins, minerals and trace elements are good for at least four to five months, so there are no performance issues if the bags were to go out of date whilst on your holding.

As a bonus, empty feed sacks will provide you with all the rubbish bags you could ever need, and any baler twine from straw bales will tie the tops closed, ready for the refuse collectors.

FEEDING GEESE

Geese are simple to feed, as the vast majority of their diet is grass. This does not mean grass cuttings collected in the grass box of your lawnmower, as these will quickly heat up and are only fit for the compost heap; geese need to graze their own supply of standing grass from a field, paddock or lawn. Geese will forage for much of the day, consuming over a kilo of grass (equating to 20 per cent of an adult sheep's intake), interspersed with bouts of preening and relaxing. Because they mostly forage for their own food, it is critical to allow the flock enough

Geese grazing.

Geese eating mixed poultry corn from a trug.

time to take in adequate feed, around seven to eight hours. Putting them on grass for only three or four hours a day is not sufficient, as they need most of the daylight hours available to them to secure enough nutrition.

With the arrival of winter and shorter daylight hours, the quality of grass will be at its poorest; this is just when the birds need plenty of nutrition to keep warm and thrive, so their diet will need supplementing. As the days shorten and throughout the winter, and possibly for some of the autumn and into early spring, depending on the quality and quantity of grass available, wheat or a mixed poultry corn should be fed as a supplementary ration at around 125–200g per day per bird (estimate at no more than 150g for every 5kg of bird bodyweight). Poultry corn is substantially wheat with a small quantity of maize in the mix, and may also include peas, calcium and grit, depending on the manufacturer. Avoid giving pure maize as this will just make your birds fat, and don't give corn to growing goslings because it doesn't have the appropriate nutrition.

To deter squirrels, rats and wild birds from stealing the daily ration of goose feed, tip it into a shallow bucket containing a few inches of fresh water. Geese are happy to dibble in water to find their wheat, and pests won't be able to get to it until the water is drunk or spilled, by which time the feed will probably have been devoured. You can't do this with poultry pellets or mash, which would turn instantly to mush and be rejected.

When broody geese come off the nest having hatched goslings, they will have eaten practically nothing but a quick daily snatch of feed for a month; in order to build up their weight, offer them mixed poultry corn for a few weeks. Their baby goslings are fed chick or duck starter crumbs for the first few weeks, and will also enjoy chopped lettuce, spinach and chard from the vegetable garden, as well as dandelion leaves.

FEEDING DUCKS

Waterfowl feed can be bought for ducks from hatch through to laying, but it can be expensive (in some cases more than double the price of chicken feed), it is not available everywhere, and in many decades of keeping waterfowl, we

A range of suitable feeders.

Duck and goose starter crumb.

have never used it. For baby birds we use chick crumbs (always the unmedicated kind), moving on to poultry grower pellets after six or so weeks. Once full grown and coming into lay, our ducks move on to poultry layer pellets, which have a higher calcium content than grower pellets.

If you prefer and can source them, duck starter crumbs are available for ducklings, providing a high energy diet for growth in the first three weeks; for those being reared for the table, duck grower or finisher pellets can be fed. Duck maintenance pellets with lower protein are designed for feeding full-grown young birds before they come into lay; at this stage these birds don't need the additional protein boost of the higher energy breeder pellet, which is appropriate once they are laying eggs.

Most manufacturers suggest that you feed birds ad lib, which, if you were being sceptical, could equate to putting out more for the birds to eat than they actually need. Certainly fast-growing youngsters need food available to them most of the time. For adults, however, give them enough for their daily requirements, but put out a less amount if there is food left at the end of the day, as you are almost certainly feeding rodents and wild birds with your generosity.

Protein Levels for Duck Feed

Duck feed	% protein	Age in weeks
Starter crumbs (also suitable for goslings)	21%	0–3
Grower/finisher pellets	17%	4–10
Maintenance pellets	15.5%	10 to onset of lay
Breeder pellets	16%	From onset of lay

DUCK

Age in weeks	0-3	4-10	10+
633CB Duck Starter Crumbs	Feed ad lib with supply of clean water		
634PB Duck Grower/Finisher Pellets		Feed ad lib through to finsh with supply of clean water	
402PB Duck Maintenance Pellets			10 weeks through to the onset of egg production. Feed ad lib with supply of clean water
401PB Duck Breeder Pellets			Feed during egg production. Feed ad lib with supply of clean water

Duck feed.

Quantities will obviously vary depending on the breed and size of the duck, but start with an estimate of 150 to 200g per adult per day, and increase or decrease as leftovers (or a lack of them) suggest. You may prefer to give half the ration in the morning and half in the afternoon if that suits your routine, but we put a day's ration in the feeder and serve it up at one go, as the birds don't gorge it all at once. In cold weather, feed consumption may need to go up, so adjust rations accordingly.

Although their frame is larger, drakes tend to stand back and let the females feed first, particularly when they are in lay. I doubt this is related to any notions of chivalry, rather that the females are simply hungrier, having put all that energy into producing an egg. Make sure that the feeders you use allow the ducks to eat simultaneously so that smaller, less confident birds don't go hungry.

If you spot any mould in the feed, do *not* give it to your birds: throw it out, clean your storage bin thoroughly, and source fresh feed. Clean out the feeders, too, if they get grubby, removing any clogged food remnants before they get damp and mouldy. It might be tempting to

Hanging feeder being used by young ducks and geese.

Treadle feeder suitable for ducks.

buy feed in bulk, but if the number of birds you have doesn't warrant this, you risk the feed going off. A month's supply is as much as you want to be storing.

Place adult bird feed at a slight distance from drinking water to minimize waste and the creation of smelly food slush. It is helpful to raise bird feeders a few inches off the ground; we hang feeders from a post, which also stops birds overturning them – a particular habit of geese – but you can put them on to raised grates or mesh that allows liquid to drain away. Avoid strewing feed directly on the ground as this creates a lot of waste and simply encourages vermin. As an alternative to hanging feeders, any heavy casseroles or pots that are no longer fit for the kitchen and can't be blown away by the wind or tipped over by the birds, make great free food troughs.

Treadle feeders are particularly useful in deterring rats, and ducks can be trained to use them; however, they are not suitable for large geese, as the combination of big feet and body size makes access to the feed something of a challenge.

For additional guidance on feeding newly hatched waterfowl, *see* Chapter 12.

SUPPLEMENTS

In our feeding regime supplements include vegetable garden and orchard treats. Waterfowl require more niacin (nicotinic acid/vitamin B3) in their diet than chickens in order to remain healthy, particularly in their early growth stages, as niacin deficiency can lead to leg and joint problems. If you notice a leg problem in young waterfowl, adding niacin can make a significant improvement. Their diet can be supplemented with liquid niacin in their drinking water, or with dried brewer's yeast or nutritional yeast flakes added to the feed; 55–70mg of niacin per kilo of feed is recommended, until birds are around ten weeks old. If using nutritional yeast, add one to two tablespoons per 200g (7oz) of feed.

Chapter 3 explained that once any food has been in a kitchen, whether domestic, professional or industrial, it is illegal for it to be fed to livestock. This includes not feeding kitchen scraps to pet poultry, or poultry not intended for meat. However, there are plenty of supplementary feedstuffs you can give to your birds if you feel the need, and are properly restrained in your offerings.

Geese and flowerbeds – don't forget to protect your dahlias!

I don't feed treats to animals, as the rations we give our birds are of themselves a complete diet. That said, a varied diet is good for everyone, so bolted lettuces, windfall apples, sprout tops, past-their-best corn cobs and other greenery from the vegetable patch and orchard are freely given. Hugely overgrown courgettes are sadly of no interest to our waterfowl, so are given to the pigs.

Give handfuls, rather than bucket- or barrowloads of green feed to your birds, or it will risk just sitting and rotting, and be yet another reason for vermin to visit.

Many people give a regular afternoon 'cream tea' equivalent of maize to their birds, which is unnecessary and fattening; but a bit of apple or greenery will keep birds busy and happy

without being gut-busting. Those who intend to dispatch and dress their own ducks and geese will soon see that the natural fat deposits in waterfowl are more than adequate, and there will be plenty of fat in a healthy bird to roast the best potatoes in the world to accompany a meal; you don't want to create excess quantities of fat by overfeeding with maize.

Not supplements precisely, but ducks in particular love a little fresh meat that they will happily source for themselves: slugs, snails, frogs, beetles, fish (if there's some in their pond) and more. Ducks will fight fiercely over a freshly caught frog – they are not dainty in their habits.

Grit

If geese can get at a builder's sack of sand, you'll know that they love it. Sand and grit enable the bird's gizzard to grind up its food effectively, so have sand and poultry grit available so that the birds can help themselves. Ducks that free range probably get enough natural grit, but if they are in pens, or if you have concerns about their ability to digest their food, or if you are finding soft-shelled eggs, do give them access to poultry grit; this is normally flint grit with oyster and other sea-shells that also offer calcium. Keep this separate from their feed so they can help themselves as necessary.

FEEDING HABITS

Greedy Ducks

Without a doubt, some duck breeds are greedier than others. Lightweight Runner and Call ducks will, of course, eat less than half the amount of the heavier breeds that weigh at least twice as much, but even on a food-to-weight ratio, Runners don't have big appetites, although they are keen foragers and devourers of slugs and other meaty delights. Heavy-breed ducks rush for the feed as soon as they are let out in the morning, while lighter breeds appear to have a more nonchalant, less desperate attitude.

Grazing for Geese

Orchards are great places to keep geese, but don't let them loose until the trees have been properly protected. A ring of chicken wire closely wrapped around a tree trunk is not adequate (and is also likely to damage the tree in time), and nor are the plastic spiral guards that are used to deter rabbits. Geese have all day in which to use their dexterous, busy bills, so you need tree guards that keep them out of reach of the trunks. You don't need to make large guards such as are designed to keep sheep at bay, but they do need to be able to exclude a long, eager, extended goose neck.

If you are hoping to harvest windfalls for eating, cooking or cider making, move the geese elsewhere until you've had your fill, and then let them back in to enjoy the remainder of the fallen fruit.

Shallow bucket – a cheap, cheerful and effective option for drinking water.

DRINKING WATER

It's worth repeating that waterfowl need clean drinking water available throughout the day. The easiest and cheapest option for adult birds is to use shallow, low-sided 15ltr buckets with handles, that can be taken to a tap, if a hose can't be taken to them. You will need at least one of these per pair of geese, and per trio of ducks. Rubber trugs are sturdier, last for ever, and are great if in reach of the hose, but they can't be carried when full of water (as you will discover if you try!). The shorter the bird, the shallower the water container needs to be, and a clean, new cat-litter tray (without the litter, naturally) is a good option for birds of short stature.

When planning your waterfowl infrastructure, consider where you will need a standpipe or two, or even more. When the weather is cold and wet and the temptation is strong to cut corners and head back inside for the comfort of your sofa, having water access in multiple places will keep you sane and save time. The same length of waterpipe can have multiple T-piece connections leading to individual taps sited

Standpipe.

Making use of rainwater.

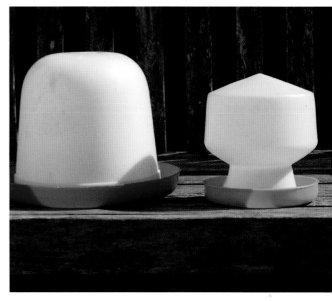

Chicken drinkers – only suitable for very young waterfowl.

where needed. When the weather is hot, you'll need to refill the buckets at least once during the day, or more simply, double the number of drinkers so that the birds don't go thirsty.

If you are lucky enough to have a natural spring, you can capture the water in a container for the birds to use, creating a permanently refreshed supply.

If you have a suitable building, a simple infrastructure can be added to harvest rainwater, collecting it via a downpipe into a water butt or even a much larger water storage tank. If the waterfowl sheds are large enough they would benefit from guttering, and the addition of a rainwater collection tank would be a satisfyingly eco-friendly solution for minimizing water bills. The lazy version is to place a water container strategically on top of a rainwater drain. Be aware that there is a potential risk of infection from wild bird faeces, so it may be a good idea to hose down roofs used for this purpose from time to time.

Although suitable for day-old ducklings and goslings, the plastic hen drinkers of the type shown are unsuitable for waterfowl over a

couple of weeks of age as they can no longer fit their bills in to get a decent drink. At this stage I move them on to a galvanized poultry drinker as the drinking area is wider and deeper. To avoid baby birds becoming waterlogged and potentially drowning, don't use deep waterbowls until birds are six to eight weeks old, and put a flat stone or brick into any larger container to allow the youngsters an island refuge, as well as a way out. By the time they are eight weeks, they will be using the same type of drinker (an open bucket or bowl) as adults.

Be aware that for young birds, the temptation to swim in their water buckets is insurmountable, so refresh these as necessary and provide more than one bucket. If there is a convenient wall or post, find a method for hanging the drinker above ground level, so that the birds can't step into it and can still drink easily.

The neat automatic drinker with a hose attachment connected to a water supply keeps water fairly clean, and ducks can't get in to swim and muck in it. Even so, you'll have to empty out any lumps of mud, gravel and other earthy detritus at the end of each day as waterfowl

can't help sifting through the earth and depositing whatever they come up with in their drinkers. The creative DIY enthusiast could make something similar with an appropriately adapted plastic container, with a ballcock and hose connector.

Having hard standing for water drinkers is ideal, particularly in permanent pens; alternatively a thick layer of pea gravel will help as long as you choose gravel that has rounded surfaces, and not sharp stones that would damage webbed feet. Inside temporary pens boggy conditions can be limited by putting down a paving slab, or at the very least moving your drinkers regularly. Bacteria and flies gather in muddy conditions, so although not avoidable entirely, minimize these breeding grounds as much as possible; it's healthier for the birds, and avoids having to refill water buckets each evening in a cloud of biting midges.

Galvanized drinker for growing birds.

OPPOSITE: *Water for commercial meat birds.*

Automatic duck drinker.

Toxic Plants

There are many plants that are poisonous to livestock, and some, such as yew, are deadly almost instantly when eaten. I'm not suggesting that you wield a flamethrower around your precious flower garden, but the plants listed below should be excluded from your bird pens. Foxgloves, vetch and meadow buttercup grow in abundance around our farm, and the geese have access to them, but I have never yet seen them tempted to eat them, possibly because they always have plenty of grass to graze. But just because they haven't in the past doesn't mean they won't in the future; you need to do your own risk assessment. It's like believing the cat and the canary are great mates until the afternoon comes when you find Ginger with a handful of yellow feathers in his mouth.

Black (deadly) nightshade
Castor bean
Corncockle
Delphinium
Fungi (not all)
Hyacinth bulbs
Laburnum seeds
Oleander
Ragwort
Rhubarb leaves
Vetch

Bracken
Clematis
Daffodil bulbs
Ferns (some)
Hemlock
Hydrangea
Lily of the valley
Potato sprouts
Rapeseed
Sweet pea
Yew

Bryony
Common St John's Wort
Daphne berries
Foxglove
Henbane
Iris
Meadow buttercup
Privet
Rhododendron
Tulip

CHAPTER NINE

Health and Welfare

KEEPING A HEALTHY FLOCK

The good news about keeping waterfowl is that they are more resistant to disease and parasites than chickens; you won't often see a poorly duck or goose squatting sadly in a corner. However, in the interests of maintaining a healthy flock it's worth summarizing points covered in more detail in other chapters. Prevention of disease is always better than dealing with the consequences, wherever possible.

If you feel you do not have sufficient knowledge with regard to the diagnosis or treatment of ailments – which is likely for all but the most experienced keeper – please do not refer to the notoriously unreliable Facebook – and other social media platforms are just as problematic – which will supply a mix of contradictory, illegal and sometimes downright dangerous disinformation. The trouble is that even the best information gets lost in the wash of nonsense on the internet.

Please talk to your vet; it's what they are there for. With their help you will build expertise and be able to determine when treatment is suitable (effective and/or economic), and when humane dispatch is the best route either to save a bird from suffering or to maintain the health of the rest of the flock. Your vet will also be essential for prescribing any prescription-only medicines (POM).

OPPOSITE: *Goose feet.*

RIGHT: *Geese in snow, Ashwell, Dolton, February 1978. James Ravilious/ Beaford Archive*

Checklist for Maintaining a Healthy Flock

The following checklist comprises a quick revisit of points covered in more detail in other chapters.

- A healthy flock starts with sourcing healthy birds or eggs.
- Always quarantine incoming birds. Where possible, do not add birds from an outside source to your own flock; if you must have additional birds, it is better to establish a second flock.
- The younger the birds, the more susceptible they are to diseases, so avoid mixing different ages.
- Choose housing that has ventilation, no draughts, and is easy to clean, and ensure there is shade available on hot days, and protection from snow and ice underfoot on cold ones.
- Keep bird huts, bedding, feeders and drinkers clean. Be particularly scrupulous about putting young birds on to clean ground that has not been recently used by adults.
- Provide a stress-free environment for your birds (avoid excessive noise, aggressive dogs and other disturbances).
- Spend time observing and handling your birds so that you notice issues before they become more serious. You want to see the following:

  - Bright eyes that are not sunken or cloudy
  - Dry nostrils
  - Smooth feathering
  - Clean feathers around the vent
  - Body neither thin nor over-fat
  - Steady breathing with no coughing, panting, sneezing or wheezing
  - Birds moving freely
  - Regular-shaped eggs produced in season

- Have a separate hut or inside space where you can isolate any bird that becomes ill.
- Feed the appropriate foodstuffs. Store feed away from vermin, and throw away any food that goes mouldy. Keep on top of pest and vermin infestations.
- Make sure there is a constant supply of clean drinking water, and water deep enough for heads to be submerged.
- Provide birds with sand and grit.
- Once a year, or more often if required, disinfect bird huts, and also clean between flocks (if starting a new flock).
- Make sure there are no items around that can cause birds harm, such as loops of baler twine, bits of plastic or broken glass, or poisonous substances such as weed killer or rat poison – or dead rodents that have been poisoned.
- Keep wild birds away from your flocks as much as possible.

Lifespan

Domestic ducks can live from eight to twelve years, although six years is more usual, particularly for larger breeds. There will always be the odd very long-lived exception – cherish them if you are lucky enough to have one of these. Geese may live up to twenty years or more, although twelve to fifteen years is more usual.

MOULTING

Moulting is an entirely natural process, so don't worry if your birds lose feathers in season. Ducks moult twice a year, geese just once. Geese lose their feathers a few weeks after they hatch their goslings, catching up with those that haven't sat,

Muscovy drake – at least twelve years old.

Goose feathers moulting after the mating season.

and which lose theirs somewhat earlier. Chickens can go practically bald when they moult, but although you will find waterfowl feathers littering your lawn or field at moulting time, the birds that have sat and hatched a brood may be slightly raggedy but they never look bald. Once new wing feathers grow, the goose will lose feathers on the body, but they are so rich in down and feather that apart from the detritus on the ground, you might not even notice. Your responsibility is to ensure they have plenty of good nutrition to enable them to regrow their feathers.

Ducks moult their wing and contour feathers, the outer protective layer of feathering, the males after the breeding season and the females prior to nesting. Just like geese that have not sat on a clutch, you have to be particularly observant to notice much difference in the look of your ducks – both just look more pristine and lovely once the moult is complete and new feathering is in place.

FIRST AID AND ASSOCIATED KIT

The following off-the shelf items will be useful:

- Purple/violet antiseptic spray for minor cuts and wounds
- Flubendazole-based wormer
- Poultry faecal worm-count pack
- Disinfectant suitable for cleaning animal housing (dry powder or liquid versions available)
- Medicated eye ointment for eye injuries and infections
- Wrap and gauze pads for bandaging injuries after treatment
- Micropore tape
- Saline solution (for wounds or eyewash)
- Scissors

Your vet will prescribe antibiotics and other prescription-only medicines as needed.

WATERFOWL ANATOMY

Familiarize yourself with waterfowl anatomy; the role and health of each part of the bird contributes to its overall fitness and wellbeing. The waterfowl bill has the combined determination of a toddler's hands and mouth – it explores everything in its range, and is crucial for feeding and foraging. The eyes are more effective than those of a human, and although the ears may not be obvious hidden under fluff or feather, they are perfectly good. The feathers are the most extraordinary part of the bird, ranging from stiff, thickly quilled flight feathers to soft layers of insulating down.

The legs and feet are particularly important as regards the bird's health, because their fitness or otherwise determines the mortality of the bird. The hock joint on a bird (often hidden under their plumage) is the equivalent of our knee, though it bends the opposite way to ours.

The oil in the preen gland is all-important for waterproofing the feathers. The gland is situated on top of the tail, hidden under the feathers (more familiarly it is the parson's nose on an oven-ready bird), and the bird rolls its head and neck backwards over the gland to stimulate oil production. The oil is then distributed all over the body as the bird preens itself.

1. Eye
2. Culmen
3. Nostril
4. Bean (or nail)
5. Upper mandible
6. Lower mandible
7. Neck feather partings
8. Breast
9. Paunch or abdomen
10. Shank
11. Claw
12. Toe
13. Web
14. Hock
15. Thigh coverts
16. Secondaries
17. Tail
18. Primaries
19. Greater coverts
20. Lesser coverts
21. Scapulars
22. Ear

External parts of the goose.

SIGNS OF ILL HEALTH

Lameness

Lameness can often be the first sign of ill health, and might indicate anything from a mechanical injury to the leg or foot (bruising, a knock, treading on something sharp), in which case bed rest and time is indicated, to intestinal worms or a virus. If your bird is lame, pick it up and check for foot and leg problems first (heat in the leg, swelling, a wound, a broken bone); if you can't find anything mechanical check its breastbone and body condition to see if it's underweight.

Lameness can be induced if birds need to step up and down, so if there is a step into/out of the goose or duck hut, make sure you provide a gentle ramp for them to use. A ramp is also helpful into a pond or sink used for bathing, and put a large stone or bricks in any pond so

1. Eye	11. Shank
2. Nape	12. Web
3. Scapular	13. Toe
4. Tertials	14. Hock
5. Secondaries	15. Keel
6. Primaries	16. Breast
7. Sex curl (drake only)	17. Bean (or nail)
8. Tail	18. Nostril
9. Undertail	19. Forehead
10. Stern	20. Crown

External parts of the duck.

birds can get out again. Heat in the leg might indicate infection, in which case antibiotics may be required.

Bumblefoot is a hard swelling on the underneath of the foot, which is normally caused by bacteria entering a small wound. Clean the site and remove any crust and pus, and treat with antibiotic spray. If the bird can spend time on water, rather than all day on land, this will take weight off the damaged foot.

Being Underweight

If a bird feels lighter than it should, and if its breastbone feels prominent (you need to feel for this, as thick down and feathers hide body condition from view), it is likely to have intestinal worms and needs worming. You want to be able to feel a well-padded breast, not a bony one. There may be other underlying problems, but effective worming should be your first step. Birds should always feel heavier

Antibiotic, antiseptic and copper/zinc skin protection sprays.

than you expect, so do pick them up from time to time so that you can gauge their weight and check their body condition.

Minor Wounds

If a bird has a minor wound, clean it and use antiseptic spray, or a spray with zinc for quick healing. If you are concerned about the potential for secondary infection, apply antibiotic spray, which is prescribed by the vet.

Breathing Problems

Gasping for breath can indicate the condition known as aspergillosis, which comes from fungal spores that develop in mouldy bedding or food. It can therefore be avoided by good management. Ensure bedding is always dry, and mucked out and changed regularly, that feeders are kept clean, and that any mouldy food is disposed of where birds can't get at it.

Although not as common in waterfowl as chickens, gape worm is possible, so treat with wormer and move birds to clean ground. Add lime to areas that have been heavily used, and leave these to rest.

An obstruction may also cause breathing problems – geese in particular are ferociously curious about everything, so keep their areas free of items they can swallow, such as any fencing staples or wire after fencing repairs.

Some breeds of duck (Call, Indian Runner, Black East Indian) can develop protruding air sacs under the eye, which makes them susceptible to mycoplasma: this manifests as a discharge and an unpleasant smell from the nostrils, sneezing, and bubbling in the corner of the eye. Antibiotic treatment will be required, given by injection or added to the drinking water.

The most common respiratory disease in ducks is caused by *Riemerella* bacteria, which invades either through inhalation or cuts on the feet. The bacteria infects the brain, joints, oviducts and the respiratory system, and more commonly affects ducklings rather than older birds. Signs of infection in ducklings are sneezing, runny nose and eyes, lameness, dullness, diarrhoea and a twisted head. Antibiotics may be prescribed but recovery is poor. Maintaining good hygiene in the duck house and run, and clean water, are the best preventative measures.

Abnormal Droppings

Normal goose droppings are tubular, and as geese have a mostly grass-based diet, are normally dark green in colour. Duck droppings can be more of a watery splat, and come in various colours and textures depending on what they are eating. Urine and faeces are expelled at the same time, the urine in bird droppings being the white, pasty liquid that usually decorates the top of the faeces.

Blood in the droppings indicates coccidiosis (*see* the section Internal Parasites below). Waterfowl can suffer from diarrhoea as a result

Normal duck poo.

Normal goose poo.

of coccidiosis, worms, and viral or bacterial infections, and if left untreated this will lead to weight loss and dehydration. Poor quality feed or inappropriate food treats can also lead to diarrhoea, as will dirty drinking water.

After providing clean, good quality feed and water, and ensuring that treats are not fed, investigate further causes by sending faecal samples to a lab for diagnosis. In addition, try to keep wild waterfowl away from your home flocks as they can transmit diseases that cause diarrhoea.

Dropped Willy or Penis Paralysis

I have only seen penis paralysis once, in a large-breed drake; it improved, but ganders can also contract this condition, in which the penis fails to retract and drags along the ground. The penis should be washed and kept clean, and the bird put on bed rest on clean straw with food and water to hand for a few weeks, to see if the organ will retract. If it doesn't it is likely to atrophy and will need removing (by a vet) to avoid it going septic and compromising the bird further. This will, of course, make the bird redundant as a breeder.

Prolapse

A prolapse is when interior organ(s) are pushed outside the body cavity; in the case of waterfowl this is likely to be the oviduct prolapsing out of the vent. This may be due to a bird being over fat, laying too large an egg, or because of a defect in the oviduct. Although the oviduct can be pushed back in and a stitch made to hold it in place, this is probably a pointless exercise in a laying bird, which will be putting constant pressure on the oviduct. In this case euthanasia is required.

Egg Bound

Young layers may have difficulty laying eggs, and if a youngster is looking hunched this may indicate that it is egg bound. Warm (not hot) vegetable olive rubbed around and just inside the cloaca (vent) may help, as does putting the bird in a dark space under a heat lamp. Gentle manipulation may help, but avoid breaking the egg inside the cloaca as damage from a broken shell can lead to blood poisoning.

Strange Eggs

Both ducks and geese can produce a variety of egg oddities, but most are nothing to worry about. A 'wind' egg is one that has been formed without a yolk; it is much smaller than usual, and occasionally it is two eggs combined. Wind eggs tend to come from young birds early in their first laying season. A double yolker is the opposite – it is an egg that contains two yolks,

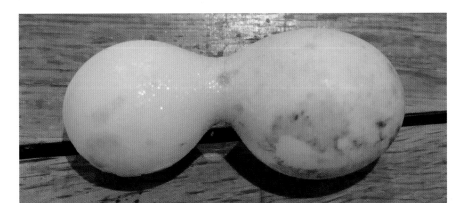

Double wind egg.

Aylesbury eggs: a normal egg and two wind eggs.

Inside a double yolker.

and is significantly larger than the norm. These are something of a breakfast treat: they are not viable, so do not attempt to hatch them. Shells that are thin or soft might be improved through a better diet, making poultry grit available, and offering calcified seaweed.

Wing Problems

Occasionally birds have minor wing deformities; although they may be unsightly, they don't seem to impede the bird. However, do not use these birds for breeding. Other problems such as angel wing – where one or both wings hang down – are linked to inappropriate levels of protein in the feed. Ensure young birds are on the correct diet, moving them on to a grower ration as soon as their feathers start to appear. If one or both wings hang down, strap them closely to the body for a week or ten days (micropore tape is useful for this), replacing the tape every couple of days.

Eye Problems and Damage

In its eagerness to mate, a drake can damage a duck's eyes by holding on to the eye area, rather than the neck feathers, as it mounts and mates. A gander also holds on to the feathers at the top of the female's head as it mates, and can make the goose's head quite bald, though temporarily; however, I have never yet seen a gander damage a goose's eye in the same way. Remove the duck from the group to allow it to recover, as it may well be the drake's favourite and will be receiving constant attention. Antibiotic eye cream may be needed; talk to your vet.

Other eye problems such as sticky eye can be exacerbated because inadequate clean water is provided for washing and preening. The eyes look gummy and sore, and antibiotic eye cream should be applied when the bird is put into its hut for the night so it doesn't immediately wash it off.

VACCINATION

The vaccination of poultry and the majority of captive birds against most diseases, including avian influenza, is not currently permitted in the UK. Only vaccination against salmonella is currently licensed for use for ducks, and nothing in geese, even though vaccines do exist for duck viral enteritis, duck viral hepatitis and goose parvovirus, and are being developed for duck parvovirus and avian flu.

There are many vaccines listed for use in chickens and 'poultry' – for example against Newcastle disease, infectious bursal disease and *E. coli*; these tend to involve small dose volumes, sometimes have complex methods of administration, and can come in pack sizes of a thousand doses! Even the largest manufacturer of poultry medication in the UK doesn't have

any licensed products for the vaccination of ducks and geese. My farm vet Chris Just BVSc MRCVS tried to set up some vaccination protocols for a small free-range egg producer and had to order tens of thousands of doses of vaccine to get a delivery from the suppliers. He has some pragmatic advice for small flock keepers:

We can expect very low disease incidence in low intensity stock keeping with good hygiene and biosecurity. That would be the area to focus on; with a nod towards the attitude that if you have a problem, work with your vet to sort a bespoke solution. Commercial duck and geese farms do use vaccines designed for poultry, but that doesn't mean it is a necessary thing to do with your backyard flock. Good health and medical practice does not have to involve drugs.

DEALING WITH PARASITES

Internal Parasites

Internal parasites are organisms such as worms that live in the body of the host, occupying the digestive tract or body cavities, body organs, blood, tissues or cells. Any bird that is thin or coughing should be wormed.

Flubendazole is the active ingredient within the most commonly used poultry wormer Flubenvet, a powder that is mixed with the appropriate amount of feed and fed to your birds for seven consecutive days. Flubenvet 1 per cent is for domestic flocks and is effective against the eggs, larvae and adult worms of the following: gapeworm (*Syngamus trachea*), large roundworm (*Ascaridia galli*), caecal worm (*Heterakis gallinarum*), hairworm (*Capillaria* spp.) and gizzard worm (*Amidostomum anseris*). Flubenvet can be obtained on veterinary prescription from your vet or a suitably qualified person (SQP), and most agricultural merchants will have their own in-house SQPs. The stronger Flubenvet 5 per cent is for commercial use, and must be incorporated into feed at a registered mill.

Treated birds may be slaughtered for human consumption only after seven days from the last treatment; there is no recommended withholding period for domestic poultry producing eggs for human consumption when used at the recommended rate. If you are concerned about residues in eggs, consider worming your birds when they are not in lay. At least two companies in the UK produce poultry feed with Flubendazole added, and as the medication is the texture and weight of talcum powder, which can get blown away in a puff of wind, this is a really good way of treating your birds. Buy a week's worth of in-feed wormer when required, and the job is done.

If necessary there are off-licence alternatives that a vet can prescribe, such as one of the ivermectins, which can be given by mouth or injected, depending on the type prescribed.

There are two useful points in the year to consider worming: as your flocks start to sit on a clutch, which minimizes the transfer of the worm burden to young hatchlings; and if a second worming is desirable, when the laying season is finished. There is advice suggesting that domestic birds should be wormed quarterly; however, rather than worming too frequently or unnecessarily, send fresh bird droppings (collected either from individuals, or as a pooled, composite sample from a group of birds) to be checked for worm counts to identify if the birds are carrying a worm burden

Poultry faecal sampling packs.

that needs treatment. Your vet can advise, or there are several laboratories that accept posted samples. Tests can check for roundworm, gizzard worm, hookworm, hairworm, caecal worm, gapeworm and coccidiosis. Labs can also test for salmonella.

Apple cider vinegar can be given to birds (diluted in water, and always served in a plastic container) as a tonic and to benefit the health of the gut, but it won't cure a worm infestation. Always provide fresh plain water alongside. Best practice is to organize a stock rotation with your birds, giving them access to fresh ground, particularly if you are heavily stocked, to prevent a build-up of worm eggs on the pasture. You can also clean waterfowl houses with a disinfectant that is licensed as effective against worm eggs, coccidial oocysts, bacteria, fungi, and viruses such as interkokask.

Coccidiosis is a disease caused by parasites that live inside the cells lining the bird's intestine, causing bleeding and swelling. Ducklings and goslings up to two months of age are at particular risk. Birds become dehydrated and cannot absorb nutrients from their food, and will soon die if left untreated. If you observe them closely you should see blood in the droppings, and affected birds will be listless. In the absence of licensed alternatives, vets will prescribe an anti-coccidial drug that is administered diluted with drinking water or as an oral drench.

External Parasites

Red Mite or Northern Mite
A healthy goose or duck that is engaging in its usual routine of washing and preening will remove most external parasites, but if they are housed with chickens and not given adequate clean water they can be at risk of red mite or Northern mite. Birds that are sitting on a clutch are more prone to infestation, as is any bird that is poorly. Red mite lives in any cracks in the bird hut (which is why shiplap, however attractive, is unsuitable for poultry housing) or under roofing felt, while Northern mite lives permanently on the bird.

Keep your huts clean with a proper disinfectant and insecticide such as Poultry Shield; diatomaceous earth can be safely used as a preventative measure against mites and lice, scattered in the bedding and on the birds. If your birds are permanently scratching themselves or losing condition, ask your vet to prescribe a pour-on insecticide (applied to the back of the neck or under the wing) or an injectable one.

Flystrike
In flystrike, blackbottle, bluebottle or greenbottle flies are attracted to dirty, faecal- or urine-stained fleece, skin, fur or feather. The flies lay eggs, which hatch as larvae within

Red Top fly trap. (www.redtop-flytraps.com)

Muscovy caruncles.

twelve hours and feed on skin and faecal material; these become mature maggots in about three days, depending on humidity and temperature, and these maggots excrete enzymes and devour the affected flesh. Although uncommon in birds, the open-feathered (rather than tight-feathered) breeds such as the Toulouse and Sebastopol geese or Pekin duck can be susceptible, and the vent area is most likely to be affected.

When the weather is warm and muggy, keep a close eye on your birds to check for irritation and early infestation. Wash off and remove any maggots and fly eggs, and if the skin is damaged, spray with antibiotic spray. To minimize flies, and if you have real problems with fly infestations, use preventative non-toxic fly-control traps around the bird huts.

OTHER DISEASES

There are many diseases that waterfowl can get, from venereal diseases to fowl cholera (carried by rats); just a few are outlined below. Many can be avoided by practising good sanitation, minimizing contact with wild waterfowl, controlling vermin, and sourcing healthy stock.

Avian Influenza (Bird Flu)

Avian influenza is a notifiable disease. Mortality and respiratory problems are less severe in waterfowl than in chickens and turkeys, but birds can die quickly without showing any signs of illness, and waterfowl are significant carriers. Signs can include the following: a lack of co-ordination, purple discoloration of caruncles and legs, a drop in egg production, soft-shelled or misshapen eggs, a lack of energy and appetite, diarrhoea, swelling of the head, eyelids, caruncles and hocks, and nasal discharge. There is no vaccination or treatment for the disease, and affected flocks are culled.

Cryptosporidiosis

Caused by parasites, this infects the lungs and intestine, and is found worldwide wherever commercial poultry are raised. There is no treatment, so good hygiene is recommended as the best preventative measure.

Duck Viral Enteritis (Duck Plague)

Duck viral enteritis (DVE) is a viral infection of ducks and geese often spread from wild waterfowl through contaminated water. Even if birds recover, they remain carriers for life. Signs of DVE include sudden death during migration of wild birds between April and June, wing walking (using their wings as crutches), runny nose and eyes, loss of appetite and bloody diarrhoea. Diagnosis is by post mortem. Antibiotics may be prescribed; there is no vaccine available for use in the UK. Where possible, prevent wild waterfowl from accessing drinking and swimming water used by your birds.

Duck Viral Hepatitis

Duck viral hepatitis is highly contagious and primarily affects the livers of young ducklings up to three weeks of age, which suffer high mortality; immunity builds with age, so that birds of four to five weeks of age suffer negligible mortality. The signs are sudden death, twisted neck and dullness; diagnosis requires a post mortem by a vet. Good hygiene and minimizing stress help to reduce the risk of this virus. Antibiotics may

be prescribed; there is no vaccine available for use in the UK.

Erysipelas

Erysipelas is an acute infection of individual young and adult birds caused by bacteria; it affects many species, including humans. Birds appear depressed, suffer diarrhoea, and die suddenly. Treatment is by antibiotics; birds that recover have a high degree of resistance to re-infection.

Goose Parvovirus (Derzy's Disease)

Other names for this virus include goose plague, goose hepatitis, goose enteritis and goose influenza. It is highly contagious, and is excreted by adults through droppings and their eggs; mortality can be total in young goslings. If goslings are four to five weeks or older when infected, the mortality rate may be negligible. There is no treatment, and imported eggs and birds should be avoided.

Newcastle Disease

This is a highly infectious viral disease with a high mortality rate, and a notifiable disease in the UK. Birds found to be suffering from highly pathogenic strains of the disease are compulsorily slaughtered. Signs include diarrhoea and respiratory problems.

Streptococcus

Streptococci are bacteria that live in the bird's gut; an underlying problem can make the bacterial load significant. It affects young birds up to three weeks old, the first signs being sudden death. Treatment is by antibiotics; prevention is through good hygiene.

GENETIC ISSUES

A crest is a feather pompom on the top of a duck's head caused by a genetic mutation. The crested gene (or more correctly, allele) can be bred into any duck with the exception

Crested duck.

of the Muscovy. A crested parent will produce a percentage of crested offspring. If just one parent has the crested gene, 50 per cent of the ducklings will be crested. If both parents are crested, 25 per cent of the ducklings will die in shell, 25 per cent will have crests, and 50 per cent will not have crests.

There are significant downsides to be aware of in crested ducks: as mentioned above, a percentage of ducklings that receive the gene from both parents will die in shell. An adult female with a large crest can easily be injured during mating; and the crest gene can cause fat in the skull, which can seriously impede a duck's mobility, cause seizures, neurological problems and early death.

POISONS

It is critical to ensure that inquisitive birds have no access to rat poisons or rats that have succumbed to being poisoned. Other things that might not be considered as being toxic to waterfowl include avocado and eucalyptus. Chopped straw and wood-based products for horse bedding are often treated with eucalyptus

Doop and sharps bins.

oil, so these are unsuitable for waterfowl bedding. Products containing by-products of the cocoa industry are also toxic, so avoid using cocoa-shell mulch on any areas of garden where your birds might stray. A list of poisonous plants is given in Chapter 8.

If you shoot vermin on your holding use non-toxic shot that does not contain lead; swallowing a single lead pellet can prove fatal to waterfowl. Shot containing steel, iron, tungsten or bismuth are available as alternatives.

BEST PRACTICE ISSUES

Isolation Space for Poorly Birds
Keep a clean hut available for use as a hospital space if required so that birds can be isolated, when undergoing treatment, or to stop the spread of infection, or to allow them bed rest where they won't be bothered by other birds.

Humane Dispatch
If a bird is suffering and beyond treatment, it should be humanely dispatched. A vet can do this for you, or you can follow the guidance from the Humane Slaughter Association, as described in Chapter 11.

Safe Disposal of Sharps and Unused Medication
Use a proper sharps bin to dispose of used sharps: hypodermic needles, syringes with needles, scalpels, blades, disposable scissors, suture equipment. A sharps bin is a specially designed lidded box that you can buy from your vet, which will include the cost of disposal. For any unused/out of date medication a Doop (Destruction of Old Pharmaceuticals) bin can be bought from your vet: this will include the cost of its disposal when full.

CHAPTER TEN

Keeping Your Flock Secure

All the care you lavish on your birds takes time, energy and money. It's not possible to eliminate all elements of risk, but there are certainly things you can and should do to help keep your birds secure. Some risks are to do with disease, others with theft or predator attacks, and your care and attention can help prevent your birds falling victim to illness and predation.

PREDATORS AND PESTS

Theft

The saddest theft is that perpetrated by humans. Rural crime seems to be permanently on the increase, and ranges from the opportunistic thief to stealing to order. What gets in the news may be the rustling of whole flocks of sheep,

OPPOSITE: *Geese and electric fence tape.*

Muscovy ducks by a woodrick, Deckport, Hatherleigh, May 1976. James Ravilious/Beaford Archive

Ducks keeping an eye on a buzzard.

tractors and quad bikes, but birds are also stolen, from a handful of hens to put in the pot to the targeting of prize breeding birds. Geese are particularly vulnerable to theft during the days that lead up to Christmas. So what can you do to keep your birds safe? The following suggestions might be useful:

- Talk to your local police rural crime team to get advice on what the particular risks are in your area, and guidance on how to secure your birds.
- Think about using technology to your advantage – video cameras can capture crucial evidence and act as a deterrent. Remote gate and intruder alarms may be appropriate for your set-up, and you might consider wildlife cameras with remote viewing. For those without an electricity supply, battery-powered wi-fi cameras are available that you can access from a smartphone.
- Put up signage that says you have CCTV, an alarm system, large guard dogs.
- Put motion sensor lighting in vulnerable areas.
- If you work from home, don't ignore excessive quacking and honking during the day (or night): it's probable that something is going on that you should check. Better safe than sorry.
- Join your neighbourhood social media groups; police and others in the area can alert you to any current issues.
- Consider taking out insurance cover.
- Keep tools and equipment secure, particularly those that could aid a break-in such as bolt croppers, ladders and crowbars.
- Don't keep precious birds off site (this can be challenging for people with livestock not kept at home).
- Where possible don't keep your birds by the boundaries of your holding.
- Take photographs of your birds; these help in social media appeals if birds are taken.
- Paint or mark batteries and energizer units with your postcode.

Dogs and Dog Walkers

Dogs that are not kept under control by their owners can hunt and kill poultry. I wish there was a simple answer to this one, but if you have a footpath running through your patch, it's almost inevitable that you will have trouble

Dogs and geese.

with dogs. For some unfathomable reason, there are still owners who can't imagine their beloved pet as a hunter with a killer instinct, and they consider the request to keep their dog on a lead as an infringement of their rights. If dog walkers come through your holding you will either have to fence the path so that dogs cannot wander off, or keep your birds contained at all times.

Containment doesn't have to mean overly restricted; although costly, large areas can be fenced for your birds, and this will double as protection from other predators such as foxes and badgers. Make sure that any electrified fence is labelled and doesn't obstruct footpaths.

Rats

Rats are the bane of every poultry keeper's life, and pose all sorts of risks to your birds, contaminating and eating their feed, stealing their eggs, killing goslings and ducklings, spreading infections such as salmonella and mycoplasma, and carrying parasites including fleas and mites. For young birds kept in a covered run, consider also putting mesh on the bottom, to give day-time protection from rats. Large rats will attack adult ducks, too; when we found a bloodied duck one morning, we set up a wildlife camera, and a huge rat was incontrovertibly seen to be the guilty party. There are various ways to control (no euphemisms intended, I mean kill) rats, but first and foremost you need to do everything you can to dissuade them from taking over. A pair of rats can become two hundred in just one year, so it doesn't take long for a few rats to become a major infestation.

Minimize the presence of rats wherever possible by taking the following precautions:

- Ensure that you keep your feed bags stored in rat-proof containers. This doesn't necessarily mean investing in expensive feed bins: metal dustbins and old chest freezers (with locks and electricity cables removed) make excellent, long-lasting rat-proof feed bins.
- Clear up any spilled feed, and always remove open feeders at night to a rat-proof container.
- Use treadle feeders for ducks.
- Keep things tidy in poultry areas so that rats don't create rat runs.
- Repair and block any holes into bird huts, and if chicken wire has been used anywhere in the coop, swap it for much sturdier galvanized weldmesh (1 x 1cm squares maximum) that a rat can't chew through. Rats can squeeze through holes of no more than 1.25cm, and mice can get through a space half that size.
- Design your hut so that it doesn't encourage rats; a thick concrete pad underneath that has an apron bigger than the hut will stop rats living below it. Use broken bottle glass below the concrete as a further deterrent. If you build the hut on stilts make sure there is enough clearance for you to access underneath it, rather than creating a perfect rat den.
- Collect eggs before rats have a chance to gorge on them.
- If you have electric power in your huts, protect the wiring in galvanized metal trunking; you don't want a rat starting a fire by eating

through the wiring.
- Screw metal sheeting to the bottom of hut doors, which will stop rats gnawing entry holes.
- Maintain a minimum metre-wide space around the perimeter of the hut that is free of long grasses and weeds so you can check for rat runs and any rodent activity.

Rat Control

There are various ways to kill rats effectively, but some, such as poison, have a serious downside in that they impact on non-target species, even when you use proper bait boxes. The last thing you want is to have poisoned, dying rats picked up as food by owls and cats, or poisoned mice being eaten by your birds. The table indicates other humane options that are open to you.

Fenn trap. (trapbarn.com)

Fenn trap set up in a protective tunnel.

Methods of Rat Control

Methods of rat control	Advice	Advantages	Disadvantages
Break back or snap trap	Place around the edges of a space, in a tunnel structure (for darkness and to stop your birds, pets, non-target species reaching inside). Bait with peanut butter	Cheap, easy to set. Good for young, small rats. Can be left permanently in place	Doesn't always kill – the rat can drag it around if partially trapped in it. You have to handle the rat to remove it from the trap
Fenn trap	A strong metal spring trap. Must be placed in a tunnel, as above. Bait with peanut butter. Can be secured to stop a rat dragging it away	Highly effective, especially against large rats. Can be left permanently in place	Be very careful when setting; they are strong enough to break your fingers. You have to handle the rat to remove it from the trap
Terriers	Either your own, or ratters with dogs will come to you. Particularly fruitful when an area is being cleared of rubbish or undergrowth	Quick, effective, kills large numbers at one time	Not a one-off solution (but then neither are the others). Can only catch those they can get at
Humane or live cage traps	Catches rats live	No harm to non-target species	You now have to dispatch the rat. Releasing it somewhere else is simply relocating the problem, and it may travel up to 8km (5 miles) to get back home
Air gun	You have to be both careful and able. Use lead-free shot	Effective if you are a good shot	It can take patience, and may be time-consuming. You only see one in ten rats
Battery-run traps	Electric shock kills the rat	Effective	Not the cheapest option
Cats	Consider giving a home to a few feral cats.	Some cats are great ratters, others less so	Your wild bird population (especially fledglings) will suffer
Gas canister-powered automatic rat trap	Get a personal recommendation before you invest	Automatic reloading after a rodent kill; no poisons involved	Very expensive, and opinions differ as to their effectiveness
Noise emitter	Ultrasonic sounds can be heard by pets and some people, not just rodents	Intended to keep rodents away from an area, therefore humane	Not seen as particularly effective
Sticky boards/glue traps	Place where your birds (and wild birds) cannot get stuck	Cheap, non-toxic, effective if baited with desirable foodstuff	Inhumane and very much seen as a last resort even by the manufacturers. Check twice a day, and any rats caught live still need to be dispatched (by you). Indiscriminate in what they catch

Use of Rodenticides

The UK Rodenticide Stewardship Scheme, run by the Campaign for Responsible Rodenticide Use (CRRU), aims to ensure that rodenticides are used correctly and in ways that minimize the exposure of wildlife and other non-target animals. Only those who are able to show they are competent in the use of professional rodenticides can purchase and use poisons applied outside buildings. There are two options for poultry keepers wishing to buy and use professional rodenticide products: become a member of a CRRU-approved farm assurance scheme, or obtain a certificate through undertaking a CRRU UK approved training course (these can be done online). The simpler option if alternatives have failed to reduce the problem is to bring in the services of a certified pest controller.

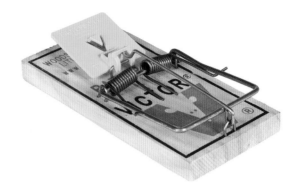

Rat snap trap.

Battery-run trap.

Rat cage trap.

Rats are neophobic (frightened of new things), so don't expect instant results; the rats will take a while to get used to new objects. Young ones are less careful and therefore more likely to be trapped first.

Peanut butter may be the number one rat bait, but other options include bacon, chocolate, nuts (particularly peanuts in their shells), apple, tomato and potato. Wear gloves when setting and removing rats from traps, the former to reduce the human smell, the latter because you don't really want to hold a dead rat with bare fingers. All traps, whatever the target species, should be checked twice a day when set.

Corvids and Other Predatory Birds

Corvids (crows, jays, rooks, magpies, jackdaws) can cause various problems for ducks and geese, stealing eggs and young birds. As described in Chapter 4, contain ducklings and goslings in covered runs. A Larsen trap, a type of cage trap, is designed to catch birds alive and unharmed, and they are used for the most part to capture magpies, which then need to be humanely dispatched. A Larsen trap can be baited with food, or with a live decoy magpie, provided welfare regulations are met. In Scotland, a cage trap must have an identifying tag obtainable from the police wildlife crime officer. It is illegal to kill ravens in the UK.

Mink

The American mink is widespread in the UK, although I've yet to see one in our patch of Devon. Although their preferred food is small mammals, including voles and rabbits, they are a risk to ducks, geese and eggs, being aggressive predators. The Wildlife Trust recommends trapping as the legally acceptable and most effective way of controlling mink, using live capture traps (cage traps), followed by licensed shooting. All traps must be fitted with an otter exclusion guard to avoid the inadvertent capture of young otters. Once a mink has been caught, it is illegal to release it back into the wild or keep it captive without a licence.

Foxes and Badgers

Far more controversial a subject than dealing with rats or the occasional magpie is managing foxes and badgers. In the case of badgers, they are a protected species and it is illegal to trap or kill them for the sake of protecting poultry. Both species are powerful diggers, so if you have any option at all, site your birds well away from any badger sett or fox earth. Both these species may be nocturnal, but there are countless incidences of birds being predated by foxes during the day.

Electrifying your runs is advisable. Create temporary movable runs using electric poultry netting for ducks; electric netting can be a

Fox hole in the perimeter fence of the duck run.

hazard for geese and I wouldn't recommend using it for them. Use the tallest netting you can find – around 1.2m (4ft) – and where the fence is placed keep the grass line trimmed in order to minimize shorting; also be sure to keep the battery charged. For permanent pens consider a double run of electric wires or tape just above ground level and 15cm (6in) above that, using insulators on the outside of the fence posts to carry the wire.

Chicken wire is fine for keeping poultry contained, but not so effective at keeping predators out as it is easily chewed through, even by rats, so use something sturdier such as wire mesh (aka hardware cloth), or weldmesh, which can be bought in panels or rolls. If you intend to bury part of the fence to create an underground barrier, use plastic-coated mesh if possible, as even galvanized wire breaks down over time.

When building permanent pens, dig a trench from the fence line, 30cm (6in) deep. Bend the mesh so that at least 30cm lies in the trench as an underground skirt, and the rest makes the bottom part of the fencing. Top with any available hardcore or concrete blocks, and backfill with the earth you've removed. It is possible to buy no-dig mesh skirts about 25cm

(5in) wide that sit on top of the ground, relying on the fact that predators will intuitively start digging at the fence edge, but if temptation and hunger are strong, these will be far more easily breached than an underground skirt. As for fence height, 2m (6 to 7ft) will suffice, ideally incorporating an outward-facing overhang.

Keep poultry huts well maintained; rotting parts are easily breached by predators, which will visit nightly to check if on this occasion they are finally able to access your birds. In general it is desirable to keep as much scrub and uncut areas as possible for the benefit of insects and other wildlife, but try to keep your poultry areas relatively clear so foxes can't lie hidden in wait for opportunities to strike.

For those with access to a sizeable piece of land, and who prefer their birds to free range, you might enjoy keeping a guard llama or alpacas: these animals have been used effectively for many years to keep predators away from all types of livestock.

If foxes are thriving in your area and encroaching regularly even where your poultry are well contained, you might decide to keep numbers down by having them shot humanely by someone trustworthy with a firearms licence, using appropriate ammunition.

Hi-tensile poultry fencing.

BIOSECURITY

Keeping things healthy means maintaining a certain level of hygiene. No one expects a goose or duck hut or run to be as clean as a hospital operating theatre or even your kitchen, but neither should it be an area of slops and faeces that welcomes maggots, flies and disease. Ducks and geese can be messy creatures, particularly when they are young and have a confined area in a brooder or youngster run. I wish you joy in trying to get ducklings to stop turning their food into sloppy mush. One helpful piece of advice is only to give them as much food as they will eat that day, so that gobs of wet food don't build up in feeders, and to keep feeders and drinkers at a distance from each other. Wash out drinkers and feeders whenever necessary, and not to a random timetable; that might mean getting out the scrubbing brush daily or weekly, as needed.

In the same way, how often you muck out bird huts doesn't necessarily have a timetable – it will depend on the size of the hut, the number of birds, the weather and the type of bedding. It's a good idea to make time to muck out weekly if that works for you, but you may find in good weather that a cursory picking out and a scattering of additional straw is all that's required, while in wet weather a thorough going through every two days with a generous layer of fresh bedding is needed. When birds emerge from their huts each morning they should be cleaner and drier than when they went to bed, particularly if they've been sploshing around in a slightly murky pond or dabbling in mud for the whole of a rainy day.

To ensure your birds keep their hut as clean as possible, put feeders and drinkers in their run or out on the grass, depending on your set-up, and not in their hut. Don't ever put feed in their night-time accommodation: instead, remove all feeders when you put birds to bed, and store them out of reach of vermin. Never put drinkers inside huts, either. People worry unnecessarily about their birds getting hungry

Biosecurity during Poultry Lockdowns

The following are UK government directives to ensure good biosecurity during poultry lockdowns. All poultry keepers are advised to take the following precautions:

- Minimize movement in and out of bird enclosures.
- Clean footwear before and after visiting birds, using an approved disinfectant at entrances and exits.
- Clean and disinfect vehicles and equipment that have come into contact with poultry.
- Keep areas where birds live clean and tidy, and regularly disinfect hard surfaces such as paths and walkways.
- Humanely control rats and mice.
- Place birds' food and water in fully enclosed areas protected from wild birds, and remove any spilled feed regularly.
- Avoid keeping ducks and geese with other poultry species, where possible.
- Keep birds separate from wildlife and wild waterfowl by putting suitable fencing around outdoor areas they access.
- Keep a close watch on birds for any signs of disease, and report any very sick birds or unexplained deaths to your vet.
- If housing is a government requirement and only small numbers of ducks and geese are involved, it may be possible to house them. If this measure is not possible, ducks and geese should be kept in fully netted areas or temporary netted structures, where practical. All feeding and watering should take place under cover.
- During an outbreak the best approach for small flocks may be to keep birds in a suitable building, such as a shed, outbuilding or garage adapted to house birds, or a new temporary structure such as a lean-to or polytunnel (note that a polytunnel will only be suitable in cooler weather). Ensure openings and ranges are netted to prevent wild birds getting in, and remove any hazardous substances.
- Ensure birds have natural light where possible, and are not kept permanently in the dark. Any artificial light should ideally follow typical day and night patterns.
- Ensure there is adequate ventilation – adult birds will tolerate low temperatures, but are very susceptible to high temperatures (temperatures should not be allowed to go significantly above 21°C).

Geese contained in their shed during an avian flu lockdown.

water will keep bedding clean and dry for much longer, and will help deter rats.

There will be times when there are specific heightened biosecurity risks, such as an avian flu outbreak; on these occasions the government will require additional protective measures for flocks of all sizes. It helps to be aware of the sort of measures likely to be put in place, so that your set-up can accommodate them with only minor tweaks. Be aware that as food and water will have to be in the hut during the day you will need to muck out more frequently.

Note that with every ensuing outbreak and subsequent lockdown, guidance is refreshed and updated, so do check current requirements. Guidance for backyard and pet keepers will be published on gov.uk, gov.scot and gov.wales.

Having a run where birds can be confined during national disease outbreaks, with fencing high enough to allow it to be covered with fruit or pheasant netting in an emergency, is a useful quarantine facility. The height means you can access the space comfortably without bending double. Fine netting gives you a safe space for birds to exercise, and where feed and drink can be placed that can't be easily accessed by wild birds. To fully avoid wild bird faeces contaminating the space, a solid roof would be required.

These disease controls tend to happen in the winter, and can last for three or four months. We have successfully moved a duck hut into the vegetable polytunnel where the ducks had a wonderful season, demolishing the last of the veg and any foolish slugs and snails. We also moved a group of utility Aylesbury ducks into the pig farrowing pen for several weeks of lockdown, and made a mesh panel with pheasant netting to put in the doorway of the goose hut. This meant we had to break our own golden rule of keeping food and water out of the hut, and so had to muck out every couple of days to keep things clean and fragrant – the smell of wet rotting food takes not much more than forty-eight hours to develop.

or thirsty during the night, but as far as humans are concerned, I don't know anyone with a raging night-time desire to keep a sandwich and a glass of milk on their bedside table just in case they wake up feeling peckish at 4am: your birds, just like you, will be asleep, and not wanting a midnight feast. Keeping huts free of food and

Ducks in our pig farrowing pen during an avian flu lockdown.

Specific measures for ducks if they need to be housed due to disease outbreaks

(Biosecurity and preventing welfare impacts in poultry and captive birds. DEFRA, 2020)

Open water sources such as troughs, filled water buckets and showers enable ducks to immerse their heads and preen effectively, keeping eyes, nostrils, beaks and plumage in a healthy condition. However, your ducks should not be given access to water that may be contaminated by wild birds. You should ensure that litter quality does not deteriorate with excess water spillage, as wet bedding can increase the spread and severity of pododermatitis (contact dermatitis on the feet) and susceptibility to other infectious diseases. Consider making use of raised, perforated plastic floors or equivalents on which to place water sources to maintain litter/straw quality. Ensure that ventilation and temperature controls are appropriate for ducks. They drink and produce more water in their droppings than other poultry, and humidity and ammonia levels can increase rapidly. Adult ducks prefer lower temperatures (around 13°C) compared to other poultry.

QUARANTINE

Part of good health management is having a clear process for quarantining any new birds that you bring on to your property. Don't just randomly put birds together and hope that because a new bird looks well that it isn't carrying a contagious disease. Birds should be quarantined from existing birds for a month, giving you time to observe any worrying symptoms. This will mean having a house and run that you can keep them in, which is not too close to birds that are already resident. When this facility is not being used to quarantine new birds, it is perfect for housing any birds that succumb to illness or injury. You should feed, muck out and attend to your original resident flock first, before you do the new birds, to avoid the risk of bringing any contamination from the new flock to the old on your clothes, boots or mucking-out tools. Keep distinct feeders and drinkers for each pen so they are not accidentally swapped over.

Bringing home young birds, or better still, hatching eggs at home, is accepted as a healthier way of introducing new stock than buying in adult birds. Adults can look perfectly healthy but they can still be carriers for disease for which your birds have not developed resistance. Keep domestic ducks and geese completely separate from other poultry species as they may not show any signs of disease but can still pass it on to chickens, turkeys and other birds.

CHAPTER ELEVEN

Raising For Meat

In times only recently past, the central focus for any book on raising livestock would be on rearing for meat. They didn't necessarily ignore keeping livestock for other purposes: these might include supporting children or vulnerable adults through care farming; breeding youngstock for selling on for others to rear; improving breed standards; producing for showing; supporting rare breeds – but the rationale that keeping livestock was primarily for food production was core. However, in many years of running smallholding courses, I have seen a step change in the reasons people give for keeping livestock. Although producing food of superior taste and quality,

with a bare inch of food miles and absolute provenance, is still regularly cited as the driving force, increasing numbers of people see their livestock primarily as pets and companions that will never sit on a plate.

We are a joyously diverse world, and for me and many the importance of quality protein in our diet is a far healthier choice both environmentally and physically than the ubiquitous ultra-processed factory foods that fill every supermarket. Goose in particular, with such a minimal requirement for hard feed, is a wonderful grass-fed meat option. Being beautifully rich foods, both ducks and geese fit perfectly into the notion of eating a little less

OPPOSITE: *Cherry Valley meat birds.*

RIGHT: *Upcott, Dolton, December 1974. Documentary photo: James Ravilious/Beaford Archive*

meat, of high quality, and making the eating of it an occasion and celebration. Even if you are lucky enough to eat it on a monthly basis, we all deserve a special meal at least that often. Chapter 14 has some of my favourite recipe suggestions for you to try.

This chapter considers the practicalities of raising birds intended for meat. If you are offended by the idea of slaughtering birds for meat, simply move on to another chapter. Even if you are keeping the ducks for eggs and for fun rather than for meat, once they are mature at sixteen to twenty weeks, I strongly recommend culling surplus drakes so that they don't compromise the wellbeing of the females; drakes do taste fabulous and will be really tender at that age.

The equipment and techniques used to dispatch birds can be harmful or lethal to humans. In no circumstances can the author or publisher accept any liability for the way in which information contained in this book is used, or for any loss, damage, death or injury caused, since this depends on circumstances wholly outside the writer's and publisher's control.

REARING DUCKS AND GEESE FOR MEAT

Whether you buy in young birds to rear for meat or breed your own, it is common practice to restrict the territory a meat bird has access to, in the belief that the more exercise a bird gets, the better developed its musculature and, so they say, the tougher the meat. However, a heavy meat-breed duck is not going to run marathons no matter how much space it is given, and geese need to roam to graze, so feel free to give your birds as much space as you can. I wouldn't raise meat birds any differently from any other types, and see muscle development as a good thing, providing flavour and texture; cooking properly will give you all the tender juiciness you need, and only old birds should be tough. However, you may choose to segregate those destined for meat – for example, a batch of surplus drakes or ganders. Feed them according to their age (*see* Chapter 8), and give them access to leafy greens such as grass, lettuce, Swiss chard and dandelion leaves.

Although chefs and home cooks alike sing the praises of goose and duck fat and its crucial role in the creation of the perfect roast potatoes,

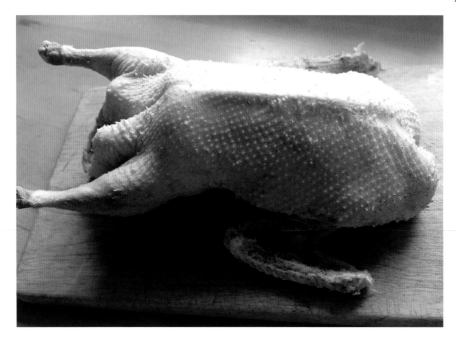

Freshly dressed duck weighing 2.07kg (4lb 9oz).

there are limits to how much fat you want from a bird. Feeding high protein/lower fat feeds works best, so choose grower or finisher rations rather than layer pellets, and don't give them bowls of maize – they don't need it, in the same way that I don't need a daily ration of chocolate éclairs. Do try different meat breeds and see what works for you, investigating beyond the usual Pekin or Cherry Valley, and raising Rouen, Saxony, Muscovy, Aylesbury and Silver Appleyard. (Cherry Valley is the leader in the modern duck production industry, and has a worldwide presence.)

For the duck-meat industry, a commercial Aylesbury/Pekin-type hybrid, including the Cherry Valley, is most commonly raised to six weeks of age, reaching more than 3.5kg (7.7lb) liveweight. White-breasted ducks are preferred as there is no risk of any dark feathering being left, which might be off-putting to the consumer. I'm not at all fussy about a little black bristle on my Berkshire pork roast or a few black quills in a parson's nose, but we all have our limits, so choose a breed according to yours.

Barbary duck (the culinary name for the Muscovy) is also available in butchers and supermarkets among their fine dining options, although the reality is that it will probably be a Muscovy crossed with a Cherry Valley-type bird. Muscovy brings some firmness and a darker, leaner aspect to the cross. Gressingham is another name you might see on the shelves and on restaurant menus, which is a Pekin/Mallard cross.

Commercially, geese are raised until they are anything from six to eight months of age, and are primarily slaughtered for the Christmas market. These geese are unlikely to be a pure breed, but for the same reason as commercially produced ducks they will be all white, such as an Embden cross or the Danish Legarth, which will produce an oven-ready bird of 6–6.5kg (13–14lb) at around six months.

For home-produced geese, any of the domestic breeds or their crosses will produce

Rouen-cross table ducks.

fine oven-ready birds, from the Roman at a maximum 4.5kg (under 10lb), which will fit in smaller ovens and feed four to six people, to the massive Embden. However, be aware that an oven-ready pure Emden may weigh as much as 10kg (22lb), which would challenge all but the biggest domestic oven, and would feed fifteen

Flock of award-winning commercial table ducks.

or more. Embdens crossed with Toulouse make a good, fast-growing, heavyweight hybrid.

The kill-out percentage (the weight of a dressed carcass as a proportion of the liveweight of the animal before slaughter) for ducks and geese is approximately 70 per cent if the neck and giblets are included.

SLAUGHTERING FOR HOME USE

Don't overface yourself by setting yourself the task of slaughtering and prepping a dozen birds in one morning. Unless you have the thumbs of Hercules and the endurance of Sisyphus, one duck or goose per person per day is more than enough to deal with. Two of us will pluck a pair of birds companionably in a morning. One of the books on my livestock shelves says

it's possible to pluck a duck in five minutes or less. I wish. To minimize the need for removing pin feathers, slaughter birds when they are in full feather and not moulting as they develop adult plumage.

The earliest age to slaughter geese is at a minimum of sixteen to eighteen weeks, and ducks at around ten weeks (for Muscovies estimate fourteen to sixteen weeks). Our preference for ducks is to wait at least six to ten weeks later when they have their adult feathering (at sixteen to twenty or more weeks). We frequently take our ducks up to twenty to twenty-four weeks, preferring to produce a bigger, slower-growing bird without stuffing them with feed, and our geese are normally raised to twenty-four to thirty-six weeks. Our older, home-grown birds are a lot less fatty than the bought ducks of my pre-farming days.

Slaughter in the morning after the birds have fasted overnight: this will minimize the risk of bird faeces contaminating the carcass as you work.

Methods of Dispatch

I strongly advise you to acquaint yourself with the *Practical Slaughter of Poultry* guidelines produced by the Humane Slaughter Association (HSA), which is available online, complete with videos and useful images. As the HSA says: 'Slaughter may never be pleasant but it can, and must, be humane. It is your responsibility to ensure that you are fully prepared in order to protect the welfare of each individual bird.' The HSA also emphasizes the moral, legal and economic reasons for best practice when slaughtering birds, explaining why it is important to treat birds well and to avoid causing them fear or pain.

It is important to remain confident, patient, calm and quiet when slaughtering birds; this means a knowledgeable confidence, not a brash, over-enthusiastic approach. If you have any doubts about your ability, you must not attempt to slaughter any bird. The HSA runs courses on humane dispatch in the UK, and to attend one would be a very valuable investment. The information below is a précis of the HSA guidelines so that you can appreciate what is involved; however, do read the full guidelines before putting them into practice.

When catching birds for slaughter use the methods described in Chapter 7, dealing with each bird individually. Once caught you will need to restrain the bird, briefly, to enable you to slaughter it quickly and efficiently. For ducks and geese an inverted cone can be used, which leaves both your hands free and confines the wings. Immediately after restraining, stunning (which causes the bird to lose consciousness) and slaughter should follow. Effective stunning can be checked with a quick touch of the bird's eye – a stunned bird will not blink when its eye is touched.

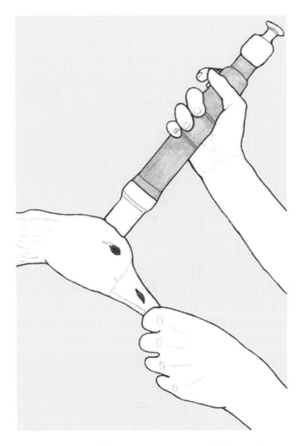

Correct positioning for concussion stunning of ducks and geese.

Immediately after stunning, within 15 seconds, the bird must be killed by either neck cutting or neck dislocation. Mechanical concussion stunning is preferable to electrical methods for ducks and geese. Two types of concussion stunning equipment are available: cartridge-powered captive-bolt stunners, and compressed air-powered captive-bolt stunners. Both come with a choice of flat or convex heads, the convex heads being suitable for ducks and geese. The captive bolt is a dangerous firearm: never point the muzzle at yourself or anyone else.

Although it is highly likely that the captive bolt stunner will also kill the bird, it is essential that they are bled within 15 seconds of stunning: cutting the neck severs both the carotid arteries or the vessels from which they arise to ensure that death occurs before consciousness can be regained. A sharp, clean knife should be used to cut across the front of the neck just below the head. To ensure food safety, birds should be kept suspended for two minutes to allow the blood to drain from the carcass before plucking and evisceration begins.

An alternative method of slaughter is by using a heavy stick or bar for birds under 5kg (11lb), although the HSA describe it as 'not ideal'. Hold the bird by the legs (and wing tips if possible), with the head and neck on the ground. An assistant should place a heavy stick (or metal bar) across the neck, behind the head. The person holding the legs should then apply firm pressure to the bar either side of the head by standing on it with both feet, either side of the bird, and immediately pull the bird's body upwards using sufficient force to dislocate the neck (this may cause some bleeding). Slaughter or killing birds by decapitation with an axe or sharp blade without prior stunning is not permitted. All birds over 5kg (11lb) must be stunned using electrical or mechanical stunning equipment, which requires the operator to have an appropriate WATOK (Welfare of Animals at the Time of Killing) licence, unless it is for home consumption.

PLUCKING – DRY AND WET

Plucking is not a job to do in a draught or outside on a windy day, as fluff, down and feather will go everywhere, including up your nose. Nor is it a job for inside the house as the mess will be prodigious. A shed or barn with a beam for hanging the birds head down, feet caught in a noose of baler twine, is the way to proceed. If you prefer to do your plucking sitting down, be prepared for being covered from head to toe in feather and down. Pluck the birds while they are still warm, so only slaughter as many birds as you intend to process immediately; once the bird is cold, plucking is harder and you are much more likely to tear the skin.

We keep extra-large cardboard boxes or outsize buckets to catch as much feather as

Geese ready for plucking.

Mrs Doris Allin plucking a goose for Christmas dinner, Cawseys, Roborough, November 1973. James Ravilious/Beaford Archive

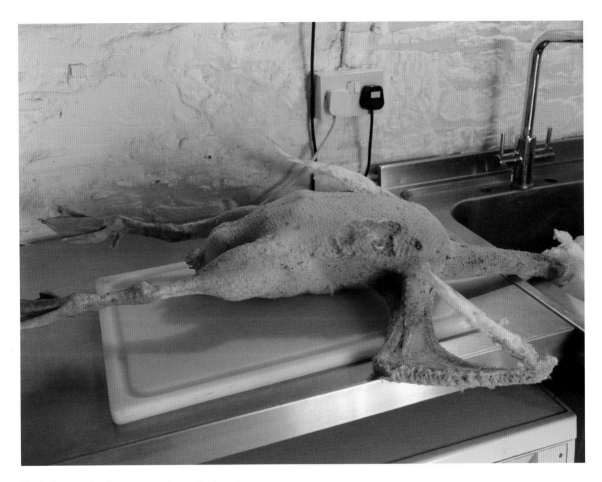

Plucked goose showing some tearing on the breast.

possible. Ducks and geese can be manually dry plucked, which gives the best quality carcass, and if you want to save the feathers, is also the method to use. It is undoubtedly a labour of love, and it takes me two hours to pluck a goose fully. The swifter alternative is wet plucking, which involves scalding the carcass by immersing it in water at 60–68°C for up to three minutes to loosen the feathers. An electric wash boiler, of the kind my mother used to boil cloth handkerchiefs, or a large tea urn no longer used to make tea, are ideal, or a very large saucepan of the kind that is found in commercial kitchens. A steamer can also be used to loosen the feather, and if you don't have a steam cleaner, a wallpaper stripper or an old steam iron both help.

To avoid tearing the skin and spoiling the look of the carcass, pull the feathers one or two at a time, not in handfuls, and pull in the direction of growth. As can be seen from the photo of the fully plucked goose, this is not always successful, but it doesn't in any way detract from the quality of the eating! Remove the wing feathers first, followed by the tail feathers, then move down the body, plucking well down the neck to the head, particularly if you want to make stuffed neck (*see* Chapter 14).

Plucking machines are efficient but expensive, and may be worth investing in if this is a regular activity (a duck would take a minute or two to be plucked by machine, geese up to ten minutes); it's certainly worth keeping a lookout for second-hand machines.

Shetland duck legs and breasts.

Once the feather has been removed it's time to start on the down and any pin feathers. You can do this either by hand, or in a wax bath where the bird is dipped into melted food-grade poultry wax, then cooled in cold water, and waxed a second time. You then pull the wax off the bird, removing the final bits of fuzz. The wax can be warmed, filtered and used again. If you've painstakingly removed the down by hand, which is what we do, it is helpful to finish off with a kitchen or workshop blowtorch: pass the flame over the bird to singe off the last vestiges of fluff and any hairy bits.

Ducks and geese could, theoretically, be skinned rather than plucked, to save you the hassle of plucking. However, in my book the crispy skin is the very best bit of the bird, so this is not something I'd contemplate. Nevertheless, when I've been given a brace of wild mallard, or have surplus small breed or runner drakes and time is tight for plucking,

the breast can be removed and the skin peeled off. The legs can also be cut from the carcass and peeled or quickly defeathered and singed, making a nice meal for two people, a leg and breast apiece. Whenever possible, though, take the time and keep the skin on the meat for its deliciousness.

GUTTING

Ducks and geese can be drawn (gutted) as soon as they have been plucked, or hung in a cool place for up to a day. This hanging – or 'ageing' – will produce a stronger, gamier flavour, which may or may not be your preference. We choose to gut our birds as soon as they are plucked. Have a bin liner or an empty feed sack close to hand for the collection and ultimately disposal of all the inedible bits. Gutting is best done on an easily disinfected surface, ideally stainless steel, next to a sink. Both plucking and gutting

Basic kit for gutting birds.

are fairly physical jobs. You'll need the following items of kitchen kit:

- Large chopping board used specifically for raw meat and poultry
- Sharp butcher's knife
- Sharp kitchen scissors
- Cloth or kitchen paper for cleaning up as you go along
- Disposable gloves if you prefer not to go in bare
- A tray for the bird
- A pot/bowl for the giblety stock items
- Rubbish sack
- Apron

You may prefer to remove the wing ends and add them to the giblet stock for gravy. We never bother removing the oil gland in the parson's nose, although some butchery guides suggest that you do. As an inveterate parson's nose eater, I enjoy that part, too.

Do *not* wash out the carcass (unless it is contaminated with faecal matter), as all this does is risk spreading bacteria all over your sink and worktop surfaces. Proper cooking, or freezing, defrosting and cooking, is all that's required. If faecal matter has been spilled on the bird, remove any gobs with kitchen paper. If you have made a worrying mess and really need to wash it, do it from the neck end downwards, and be sure to cook the bird thoroughly. You will need to be particularly scrupulous in disinfecting the sink and surrounding area afterwards.

Cool the bird properly in a cold room before putting it in the fridge or bagging it for the freezer. If you're freezing the bird, put the giblets inside the cavity, then label the bird with the weight, date and contents before putting it in the freezer. The Food Standard Agency recommends that whole birds should be frozen for no longer than a year, and

The Gutting Process

Remove the feet
Cut through the hock joint above the feet with a sharp knife or poultry shears.

Remove the head
Cut all the way through the neck as close to the head as possible

with a sharp knife or poultry shears. For geese, remove the end wing joint and discard (you may choose to do this for ducks too, as there's no meat worthy of note on this bit).

Extract the neck
You'll want the neck for giblet gravy and the neck skin for stuffing,

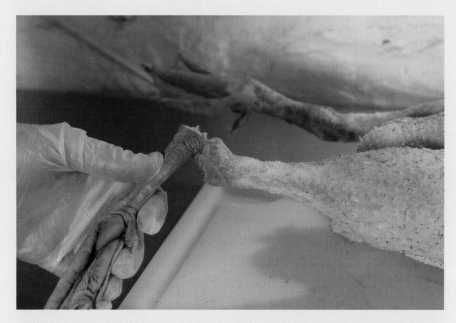

Removing the feet.

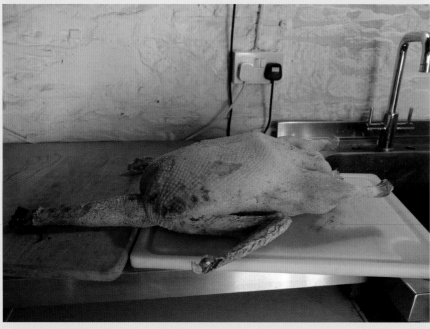

Head, feet and wing tips removed.

so cut the neck, skin and all, close to the body and then roll the skin off the neck. Remove the windpipe and oesophagus attached to the neck and discard. Put the neck and skin aside.

If the bird is empty, you can ignore the duck's crop (geese don't have one), but if it has food in – this will be obvious in the neck cavity – use your fingers to loosen the crop from the surrounding membranes attaching it to the body, and cut it off as low down in the cavity as you can get. Discard.

(continued overleaf...)

Cutting off the neck.

Peeling the skin away from the neck as a tube.

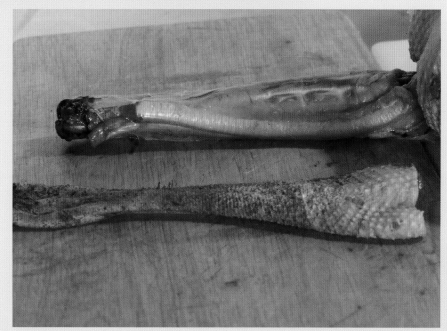

Neck and neck skin separated.

Remove the innards

Make an incision all around the vent, angling the blade away from the intestines so you don't cut into them. Hold this 'plug' and pull it gently away from the body with the intestines attached. Now run your fingers inside and around the body cavity to break any connecting membranes and to loosen any fat deposits (you may wish to keep the fat); the membranes are quite tough. Don't grab and pull at the intestines as they will break. Hook your fingers up and over the gizzard (a hard muscular ball) and then pull – the intestines will come out in an unbroken whole. Push this mass to one side away from the bird. Go back inside to extract the liver. Attached to the liver is the gall bladder, a green, broad bean-sized organ – do not break this as it contains bile – and put carefully to one side.

Go back in again to remove the heart and put aside. Check one final time for any odd bits of tubing such as the windpipe, and discard. Be aware that the windpipe is connected to the voicebox (syrinx), and pulling on it may make the carcass squawk (don't panic – the bird is definitely dead!).

Making the incision around the vent.

It is necessary to get well inside to remove the intestines.

Intestines removed.

Offal

Hold the liver carefully in one hand, gall bladder facing down, and with the other hand cut the gall bladder away from the liver, cutting into the liver, not the bladder, so there is no danger of spilling the bile. Discard the gall bladder.

Take the gizzard and cut the intestines away and discard them. Scrape any membrane and surplus fat off the gizzard, and cut round the seam leaving a hinge. Open it up and you'll find grit and possibly grass, which you discard. Peel away the thick yellow lining and discard that, too.

The lungs can be removed or left inside the cavity; when the bird is cooked this is a treat for dogs or cats.

The gizzard, liver, heart and neck are now ready to go in the stock pot or to be bagged up for the freezer. The neck skin is ready for stitching and stuffing or freezing.

Cutting the gall bladder away from the liver.

The gizzard sliced open.

Goose heart, liver, gizzard, neck and neck skin ready for cooking.

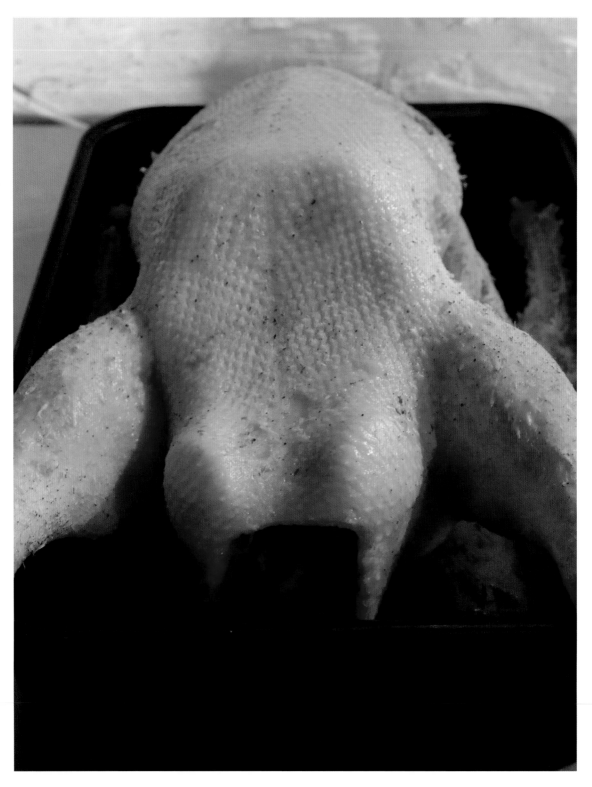

Oven-ready Pilgrim goose weighing 4kg (8.8lb).

Duck offal.

pieces for no longer than nine months before defrosting, cooking and eating; however, we have had birds in the freezer for years for personal consumption, and as long as the bag isn't ripped, which causes freezer burn (unsightly but not harmful), the birds come out as good as when they went in.

POULTRY ABATTOIRS AND HOME SLAUGHTER

It is increasingly difficult but not impossible to find poultry abattoirs willing to take a few birds to kill and dress for private customers, depending where you live. In addition, you may be able to find a licensed slaughterer who will come to your holding and slaughter birds for your own use. If those birds are for sale, they must have been raised on your holding, and must number under 10,000. Keep up to date with any changes in the law by visiting the Food Standards Agency website (or the national advisory body in your location) and searching for 'Home slaughter of livestock'.

The Small Poultry Abattoir

Anyone carrying out slaughter operations for poultry that is for sale or use beyond the immediate family must hold a Certificate of Competence, under the WATOK Regulations (refer to the Food Standards Agency Poultry Farming Guidance, which explains the rules for on-farm slaughter). Please note that it takes time to get certified, and you should plan to take the training to achieve certification by September at the latest if you intend to slaughter for the Christmas market. For information on poultry slaughter training, contact the Humane Slaughter Association.

Exemption Allowing Poultry to be Sold if Slaughtered on your Holding

At the time of writing there is an exemption (from being a licensed abattoir), which allows you to slaughter on your holding *small* quantities of poultry reared on site, and to place meat on the market for human consumption, directly to the final consumer and to *local* retail establishments, subject to certain conditions.

Producers must be registered and comply with the labelling and record-keeping requirements of Food Hygiene Regulations (England) 2006, or the equivalent legislation in Scotland, Wales and Northern Ireland. The UK interprets the terms 'small' and 'local' as follows:

> *'Small' supply is interpreted as (i) under 10,000 birds reared on the premises, or (ii) producers annually slaughtering over 10,000 birds who are members of an appropriate assurance scheme, and who either (a) dry pluck by hand, or (b) slaughter for 40 days per year or less;*
> *'Local' supply is interpreted as being the same as 'localized' (within the supplying establishment's own county plus the greater of either the neighbouring county or counties or 50km/30 miles from the boundary of the supplying establishment's county); additionally, anywhere within the UK in the two weeks preceding Christmas or Easter and for geese, Michaelmas (late September).*

For information on the legal side of selling your meat, please refer to Chapter 3.

MARKETING YOUR MEAT

There are many hefty manuals and whole degree courses on marketing, so I'll keep this limited to some very simple truths.

- If you dislike even the notion of talking about your meat to prospective customers, don't even contemplate selling it – keep to producing for your own table.
- No one will buy your produce if they don't know it exists. You have to get the message out there.
- Reaching your potential customers may cost you very little money if you are shrewd.
- A good use of social media, a genuinely attractive, informative and easy-to-find website, and an ability to share your story (people are very keen to understand provenance and animal welfare) will be more effective than paid advertising.
- A stall at your local farmers' market can be a great help in getting your product known and sold; just make sure it always looks fantastic and good enough to eat! This brings us on to the product itself, because your meat needs to look and smell great, be butchered tidily and presented well, and should delight the customers when they cook and eat it.
- Whether selling on-line or face to face, make sure your prices are obvious to every prospective customer. If they can't see the price of something, or need to phone or email you to find it out, you've lost them.
- Make sure your pricing gives you a return. Understand your costs, know what competitors are charging, and don't subsidize other people's suppers. This is a premium product – sell it as such. A business that makes a loss on every bird it sells is unsustainable.
- You have to love your own produce. People who can enthuse about what they produce are fun to talk to, informative and convincing. People love to buy direct from the farmer.
- Think about your packaging, and keep it simple, attractive and as environmentally friendly as possible. Fancy packaging makes people think they are paying over the odds.
- Recipes are always welcomed by customers; add some inspiration to their meals.

Helpful Hints for Setting up as a Small-Scale Poultry Processor
(Jade Stock, Out and About Poultry)

At the time of writing, the process below is correct for registering to slaughter and sell seasonal poultry from the farm gate. If you wish to slaughter poultry year round, or for more than forty days a year, there are different processes to be followed, including registration as a slaughterhouse and authorization from the Food Standards Agency.

- To be able to slaughter on farm for the production of poultry for sale you need to hold a WATOK ('welfare at time of killing') licence. Complete and return the on-line form available from www.gov.uk to your local Animal and Plant Health Agency (APHA) with the appropriate payment.
- Book yourself on to a course to gain a Certificate of Competence for the slaughter of poultry; this is offered by the Humane Slaughter Association.
- A vet will call and make an appointment to come and assess your process for slaughtering your birds.
- You are going to need equipment: an electric stunner and means for restricting the bird, most commonly a killing cone: a large tripod with a cone mounted on it that allows the bird to be placed into it head first so it is restrained ready for stunning and slaughter/ bleeding. Your stunner needs adjustable prongs so it can be adjusted for the size of the bird, and visible dials showing the voltage being passed through the bird. You also need a very sharp knife to make the cut to the bird's throat quickly and efficiently. Once you have these items you are ready for your vet's assessment.
- Register as a food business with your local council; this will trigger Environmental Health and Trading Standards to get in touch. Environmental Health will visit your premises to conduct an assessment, and Trading Standards will provide you with information on the legal requirements for weighing and labelling your products.
- Book yourself on a Food Hygiene Level Two course; these courses are available both face to face and online.
- You will need to write a Hazardous Analysis and Critical Control Point (HACCP) plan. This sets out how to manage the food hygiene and safety procedures in your food business, including what could go wrong and how you will minimize any risks with regard to food safety.
- Public liability insurance is an absolute must if you are selling poultry to the public, and you may want to add employer's liability cover if you have employees, also insurance for your poultry once it is in your fridges, loss of stock from the field, and possible losses due to avian influenza.

CHAPTER TWELVE

Breeding and Rearing

Eggs: little miracles. Warm in the hand, smooth, made to fit in a human palm. The continuing delight of collecting warm, freshly laid eggs is one of the most enduringly joyful of human experiences. And of course, the egg is full of potential for hatching birds naturally under the mother or a surrogate, or artificially in an incubator. For those who aren't sure, not all eggs are fertile, and unless female birds are running with a male, they will never produce a fertile egg – though they will still provide wonderful eggs for your breakfast or any other meal.

It is certainly easier hatching and rearing ducklings and goslings under a bird, but it's not always possible or advisable. Some birds don't go broody, and even if they do they can be clumsy and break everything, and you may not have an alternative, more reliable broody

OPPOSITE: *Ducklings hatched.*   ABOVE: *Millhams, Dolton, July 1986. Documentary photo: James Ravilious/Beaford Archive*

Structure of the Bird Egg

Bird eggs have a porous shell, covered in a chalky layer of calcium. When laid, a moist layer is produced, and this cuticula quickly dries, giving the egg further protection. An air sac is at the fat end of the egg – you can sometimes see this by eye through duck eggs, which have a thinner shell than those of a goose. The egg white or albumen surrounds the yolk, which is held in place by the chalazae, twisted cords that wind one way and then the other as the egg is turned in the nest (or incubator). This is why eggs in a manual incubator must be turned one way and then the other to avoid internal tearing. The yolk is surrounded by the vitelline membrane, made up of proteins. The blastodisc or germinal disc at the top of the yolk (a small white spot that you can see when you crack open an egg) is the point at which the bird will develop.

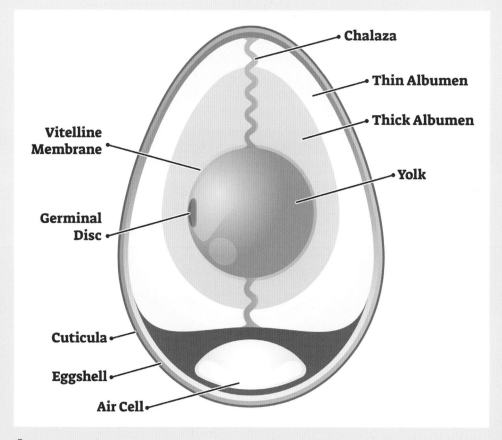

Egg structure.

Indian Runner duck eggs.

In the hatchery.

Pilgrim egg being laid.

Roman geese with goslings.

available, whether chicken, duck or goose. Hens can make great broodies for waterfowl, and they will happily sit the extra days required.

There is something very pleasing when birds come into lay. You may be thrilled that your own boiled egg and soldiers is back on the menu, and/or that the hatching season is just round the corner.

Timings for coming into lay vary hugely depending on location, weather, breed, and the vagaries of nature; I start hoping for duck eggs from February, but might be lucky from New Years' day onwards, or not until March. Traditionally, geese come into lay on Valentine's day, but as that's a human rather than an anserine construct, your geese may or may not oblige. By the end of February or early March, things do tend to get going.

Although most birds lay their eggs in the morning, and fresh laid eggs are likely to await you on opening the hut doors in the morning, plenty of our geese have laid at any time of day they see fit. Some of them seem to hold in the egg until they can leave the hut and go to their appointed 'special place' under a hedge, in the flowerbed, or wherever. I've often followed a

Goose nest in the flowerbed.

bird (at a distance) to see her disappear beneath some brambles into a makeshift nest and pop out a glossy egg almost immediately. These I remove once the goose has left, and put in a rack, ready for the incubator.

Depending on the time of year, you may start getting duck eggs from birds as young as twenty to twenty-five weeks of age. Geese will start laying in their first spring, so will be coming up to a year old.

NATURAL BREEDING: GEESE

The Laying Season

The goose laying season is short, from mid-February at the earliest (and often weeks later), to the end of May, or possibly into June if you are lucky. Occasionally a bird will lay another small clutch as late as September, which may or may not be fertile; nature does love to play with us. Fertility can be erratic, even when eggs are being laid regularly and mating is taking place. Eggs from young birds in their first laying season produce eggs from which goslings can hatch successfully, but will be smaller than those hatched from eggs from mature geese. The resulting birds may not produce the best breeding birds, although they will still make lovely pets or good meat. If you want to breed future quality breeding stock your best option is with eggs from birds in their second or later year of laying, although the corollary is also true – that birds become less fertile as they age.

The first eggs are often laid days apart, but once a goose has come properly into lay, she will produce an egg every other day. She will also give you some warning, by starting to build a visible nest. If there is a nice deep straw bed in her hut she'll use that, creating a hollow and building a surrounding dam around herself so that the eggs can't roll away, a snug place to spend most of the following month. If there is not much of a bed in the goose hut, she'll collect dead leaves, bits of twigs and any other sort of litter, natural or otherwise, that she can find; however, for hygiene and comfort it is best to provide her with straw to make a nest. As the days go by and the clutch of eggs increases in number, she will start to contemplate sitting and will pluck down from her breast to line the nest: a goose-down duvet for mum, eggs and goslings-to-be.

If a goose finds the hut less than conducive for laying – perhaps a more dominant female has taken control of the nesting site, or there are

Nene geese on the nest.

Goose egg in a drain outlet.

Pilgrim geese sharing a nest.

rats making a nuisance of themselves – anywhere that offers bowl-like possibilities will do, such as the nest created in the drain outlet (*see* picture). This is clearly unsuitable for brooding, and the goose will come off her inadequate nest each evening as birds are put away for the night, leaving the eggs to get cold or as welcome food for predators. Your options here are either to create an alternative safe indoor place for the goose to lay if you intend to let her sit, or to remove the eggs daily and hatch them in an incubator.

Although not uncommon, it is not ideal for birds to share a nest. The chances of them stealing each other's eggs are high, and eggs are overly disturbed and can get cold. Collectively geese will lay far more eggs than they can sit on effectively, although you can, of course, remove excess ones – but the chance of both sitting for the full incubation period is low.

Natural Incubation

One goose can triumphantly produce more eggs than she can properly sit on, and if you allow her to do so, this risks the viability of some or all of the eggs, as she won't be able to keep all of them adequately warm under her. Once she has started to sit, remove the grubbiest (oldest) eggs and leave her with a number she can cope with – around eight to twelve. Remove the surplus eggs when she hops off the nest for a snatch of grass, to relieve herself and gulp some water; this may happen just once a day for anything from a meagre few minutes to half an hour or more, so vigilance is required. Take a good look at her when she's sitting, and if you can see that she is leaving eggs uncovered, take more eggs away. Far better to have a few successfully hatch than a large number fail.

Do note that not every goose will sit and go broody, in which case natural incubation isn't even an option unless you use a surrogate bird. Heavy breeds are less likely to go broody, and bizarrely are sometimes too heavy for their own eggs, which break. Some geese will refuse to get off their nest; if this is the case you will need to lift the goose off once a day and put her outside so she can defecate, eat, drink and bathe. Shut the door of her hut so she can't rush straight back in without attending to her own needs, and after half an hour let her back in to get back to her sitting.

Broody hens make good surrogates for hatching goose eggs, as do Muscovy or other broody meat-breed ducks. A small broody hen could theoretically manage two or three goose eggs, although you may have to turn the goose eggs manually if she finds this challenging – so unless you have the memory of an elephant and the spare time of a sloth, I'd choose instead a meat-breed duck or large-breed hen that can cope with five or even seven eggs, depending on her size, plus the considerable upside of being able to turn the eggs herself.

Make sure fresh water and poultry corn are close to the nesting goose – not inside the hut, but close enough to it so she can keep a watchful eye on her eggs as she refreshes herself. Sitting geese lose a lot of weight during their month on the nest, so make sure they have fresh rations available in those precious short moments when they come off the nest, and that it hasn't been eaten by other birds that share the area. If this becomes a problem you'll need to split off the non-nesting birds into another pen. The gander will be very protective and will also lose weight while on guard, although not as dramatically as his nesting female.

Goose eggs take on average thirty days to incubate, but this can be from twenty-eight to thirty-four days, although I have found thirty to thirty-two days to be the norm. Incubation doesn't start from the day the egg was laid, but when the goose starts to sit solidly on her clutch of eggs, so all goslings should hatch over a period of a couple of days although there can be later ones; these late eggs may be abandoned by their mother before they get a chance to hatch if older goslings are already exploring away from the nest.

Hatching

When eggs are a day or three from hatching you can hear the goslings start to cheep. Once they've pipped a hole in the egg they work their way round, removing a lid of eggshell; the goose may help in this task. If nothing has hatched by day thirty-five, it's highly unlikely anything will. Remove the eggs from the nest before they burst – the stink of rotten goose egg is quite unpleasant and lingering. Before discarding, if the eggs are very warm, or you can hear cheeping, or see signs of pipping, do candle the eggs, just in case you've miscalculated the day that incubation began.

Once hatched, the gander is likely to take over the main parenting role while the female starts to make up for many a lost meal, sheds her dried out, lacklustre feathers, and starts to regain condition. Make sure she has plenty of feed available to her while baby goslings are given chick or duck starter crumbs for the first few weeks, and some finely chopped lettuce direct from the vegetable garden.

(*See* Chapter 8 for more information on the feeding of birds.)

If you allow and encourage geese to hatch naturally, the goose hut where she is nesting and any attached run must be safe from predation. Rats, weasels, stoats, buzzards, magpies, mink and more, that would not attempt to attack a goose, will take eggs and goslings up to several weeks of age.

NATURAL BREEDING: DUCKS

The process for natural incubation for ducks is very similar to that of geese, apart from their incubation period, which is twenty-eight days (or an extravagant thirty-five days if it's a Muscovy), but experience tells me this can be twenty-six to thirty-one days within a single batch. Meat ducks are far more likely to go broody than egg breeds, although they are not entirely reliable.

Rather than building a comprehensive and obvious nest, ducks simply create a hollow in their bedding in which to drop their eggs. Ducks commonly use each other's nest hollows,

and a morning hunt for eggs will reveal several in one place. However, they are just as likely to drop them anywhere, including on their walk from hut to run. If you are following behind at the time and pick up the egg it will be hot and wet, the dampness drying off swiftly, helping to form a protective barrier.

The Laying Season

The length of the laying season varies considerably for ducks, depending on the prolificacy of the bird – this may be from twenty to over three hundred eggs a year, so for some breeds, such as the Campbells, you'll be getting eggs daily for most of the year, while for others they'll start to lay in the spring, and will lay most days until they've reached the end of their annual capacity. My Aylesbury ducks start in February or March and stop in July, while young Shetlands can lay throughout much of the winter. (*Check* Chapter 6 for details on expected egg numbers for each breed.)

For the broodier breeds, once a clutch of eggs has been laid the duck will pluck down from her breast to create a cosy nest that retains

Khaki Campbell ducklings.

Red Shoveler ducks sitting on their adjacent nests in patches of rush.

heat for her eggs. Ducks are just as likely to choose unsuitably vulnerable nesting areas under hedges, shrubs, in clumps of nettles or anywhere else. This puts both herself and her clutch at risk of predation, so do move her to a more suitable environment. A small hut just for her is ideal, mimicking the peace and isolation she was clearly seeking out.

Natural Incubation

Like the goose, a duck can create a clutch of eggs larger than she can effectively sit on, so human intervention is required unless you want to risk the viability of some or all of the eggs. When she starts to sit, remove the dirty, oldest eggs and leave her a number she can spread herself over. A duck can sit on eight to fourteen eggs, although a big Muscovy may sit on as many as twenty. Use observation and good sense by removing surplus eggs that go beyond what your bird can manage, when she hops off the nest for a quick feed, to relieve herself and drink some water.

Make sure fresh water and feed are close to the nesting duck (not inside the hut, but close enough to it so she can keep watch over her eggs as she takes her brief break). The drake is a little less parental than a gander, but depending on the number of other females in the flock, may be more or less of a protector for a sitting female.

Incubation doesn't start from the day the egg was laid, but when the duck starts to sit firmly on her clutch of eggs, so all ducklings should hatch over a period of a couple of days. At around day twenty-five to twenty-six you will hear cheeping from the eggs, and in a couple of days the ducklings will start to appear. Make sure you remove any large water containers so that the ducklings cannot drown, putting out a suitable drinker so that they and their mother can drink. The mother can happily eat chick or duck starter crumbs put out for the baby birds, as well as adult feed raised off the ground so that only she can reach it. (*See* Chapter 8 for more information on the feeding of birds.)

In the same way as for goslings, if you allow ducklings to hatch naturally, the hut where the duck and her young are nesting and any attached run must be safe from predation.

ARTIFICIAL INCUBATION

Is there a bird-related topic that is more contentious than that of how to go about artificial incubation? There are huge disagreements as to how this should be done, and just as many individual quirks and approaches for doing something this way or that. I offer you my way of doing things, and leave others to contradict each other over correct humidity levels, surgical intervening in the final hatch, sprinkling the tears of unicorns on the eggs twice daily, and a myriad of other sub-topics. If you are eager for more of the science on good incubation practice, I thoroughly recommend Chris Ashton's book *Keeping Geese* – but the information below will get you going.

Artificial Incubation

Artificial breeding equipment:

Candler	Artificial light source used to check what's happening inside the egg.
Incubator	A controlled environment that will develop and hatch fertile eggs artificially. Many have automatic turning.
Hatcher	A controlled environment that will hatch fertile eggs artificially. It can be used for the whole incubation period, but is often preferred just for the last couple of days of incubation and for hatching, to avoid bacteria build-up in an incubator. It doesn't turn eggs.
Brooder	A brooder is simply a safe area or structure where newly hatched birds can be fed, watered, kept warm and generally cared for in the first three to four weeks of life.
Heat source	Artificial heat source to keep young birds at the correct temperature while in the brooder.

Runner duckling.

Tray of duck eggs.

Aylesbury duck double-yolker with rough shell, unsuitable for incubating.

If you are collecting your own eggs for incubation, make sure the birds have plenty of clean bedding to lay in. Eggs should be clean, well shaped, free of dings and cracks, not double yolkers, and of average size for the breed. Avoid anything that's oddly shaped, wrinkled, miniature or huge. I store eggs awaiting incubation on an egg tray, pointed side down, and put a piece of wood underneath one side of the tray so that both it and the eggs are on a small tilt. I move the wood to the opposite side, alternating each day, so that the membranes inside the shell don't stick. At all times from collection to hatching, eggs should be handled gently.

Use any dirty eggs for eating (after a judicious wash). If eggs are slightly grubby, gently brush off any dirt with a clean scouring pad; if very dirty, don't use them for hatching, although if they are precious and you are determined to use them, wash them in warm (not hot) water with egg sanitizer or human baby-bottle sterilizing fluid or tablets. The natural cuticle that forms on the outside of the shell helps keep bacteria out, and washing removes that protection. Bacteria will penetrate a very dirty shell, and the last thing you want is that bacteria multiplying in the warm environs of the incubator, causing things to go rotten and explode.

Put the eggs in the incubator at the same time, not individually as they are laid. I happily incubate eggs up to ten days old, by which time I'll have collected enough to fill an incubator. Incubating waterfowl eggs is trickier than chicken eggs – even an experienced hobby hatcher will expect no more than a 50–70 per cent hatching rate, and with some breeds even that is rather too optimistic.

Types of Incubator

Incubators have become increasingly complex with still air or fan assisted (aka forced air) options, manual and automatic turning, programmable automatic humidity control, digital displays, alarms and more. Choose according to your need and pocket. If the hatching bug firmly grabs you, you can always move on from a budget option to something fancier. In fact, having more than one incubator is very helpful, so that starter choice may come in handy as a future hatcher. I move my eggs from the incubator to a hatcher two days before they are due to hatch. The hatcher doesn't have any moving parts and is easy to clean, leaving the incubator a lot less grubby than it would otherwise be; a quick clean with incubator disinfectant and it's ready to receive the next lot of eggs.

Incubator suitable for duck eggs.

Incubator suitable for duck and goose eggs.

ABOVE: *Hatcher.*

RIGHT: *Inside the hatcher.*

Choose an incubator that matches the number of eggs you want to incubate at any one time. They are available from three hen eggs (which may fit two duck or one goose egg depending on design), all the way to huge commercial incubators for thousands of eggs. A twenty hen egg incubator with moveable separators manages fifteen duck eggs or nine goose eggs, and a forty hen egg incubator has capacity for thirty duck eggs and eighteen goose eggs. Personally, I have no patience with manual turning incubators – I would never remember to turn eggs three to five times a day, and I have other ways of filling my time – so have always plumped for automatic turning, which turns the eggs once an hour.

Candling

Candling is simply the process of using artificial light to check what's happening inside the egg, and is an important process for successful incubation. It enables you to check eggs for any cracks or defects before you put them into the incubator, to remove any non-fertile eggs from the incubator from five to seven days from starting (or from underneath a broody), to see if fertile eggs are developing as they should, and to identify any eggs that were fertile but are no longer viable. Checking for and removing unviable eggs is critical as the bacteria they give off can kill off viable

eggs, and create a vile mess and stink in your incubator if they explode.

Candlers are not expensive, and you can use a makeshift version with a small torch and a cardboard tube from inside a toilet roll if doing a one-off hatching. I've found the more powerful candlers helpful for goose eggs, which have thick shells, further improved with the addition of an egg scope that shuts off all other light sources. After two weeks in the incubator the egg contents get darker and it's challenging to see much detail in a goose egg until the bill pops into the air sac a couple of days before hatching.

ABOVE LEFT:
Home-made candler.

ABOVE RIGHT:
Mains candler.

RIGHT:
Battery egg candler with egg scope.

Progress inside Candled Eggs

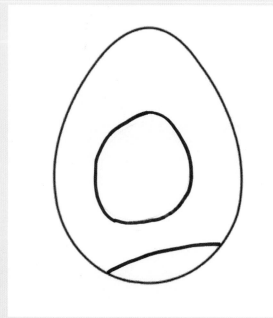

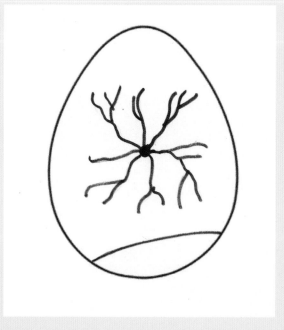

Clear egg. Candled after five to seven days: showing clear means that the egg is infertile.

Fertile egg. Candled after five to seven days: showing a blood spot and radiating blood vessels, which indicates a fertile egg.

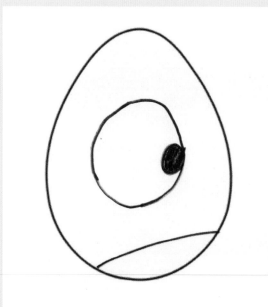

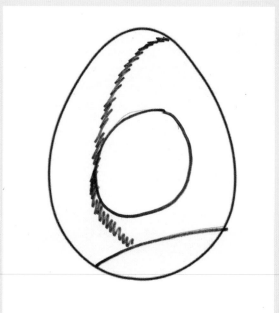

Black spot. Candled after five to seven days (or later during incubation): showing a black spot means early death, so the egg is no longer viable.

Red ring. Candled after five to seven days (or later during incubation): showing a red ring means early death, so the egg is no longer viable.

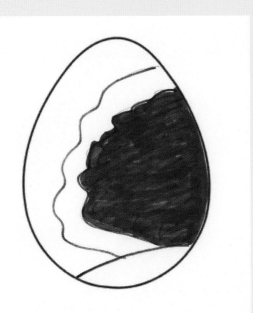

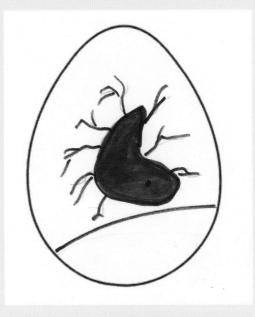

Decomposing egg. Egg showing sloppy, mobile, decomposing contents, which means the egg is no longer viable.

Developing egg. After twelve to fourteen days the egg contents of viable eggs get darker and the air sac is more defined. The contents may now be too dark to see much detail.

Further development. At around twenty-one days the air sac is bigger and the embryo contents even darker.

Close to hatching. Two days before hatching, you can see the bill in the air sack as hatching approaches.

THE INCUBATION PROCESS

If you get eggs from another breeder they will probably come with plenty of advice, such as to store them overnight in a cool room, pointed side down, then bring them up to room temperature before putting them into the incubator. I ignore this, and once the incubator is up to heat and I've checked the eggs are at room temperature and not cracked (candling will reveal hairline cracks), they go straight into the incubator. If you want to mark them in any way (for example, noting the breed, the specific bird that laid them, or the date), write on the eggs in pencil. Duck eggs can be laid on their side or pointed end down, and goose eggs should be laid on their side.

Your spotlessly clean and disinfected incubator should be brought up to 37.2–37.4°C (ideally 37.2°C for geese and 37.3°C for ducks), and then you have two choices: wet or dry incubation. For wet incubation put plenty of water in the channels provided and keep this topped up every couple of days or so, which means adding warm water so as not to reduce the temperature. Do follow the instructions that come with your incubator, as every make has its quirks. I am a convert to dry incubation and don't put any water in the

incubator channels at all until the eggs start pipping. I find that in the humid environment of Devon, duck and goose eggs lose the amount of water required for hatching far more effectively with dry incubation.

Don't cram the eggs in too tightly, or any dividers in the incubator might crack some of the shells. Do check each day that the incubator is working: if a fuse blows or you have a power cut, all your effort can go for nothing if you don't deal with it within a couple of hours. A mains failure alarm is invaluable. I keep my incubators in a cool, dark room with a steady temperature, but out of necessity have also hatched in a sunny kitchen; keeping the internal temperature steady is key to success.

At five to seven days the eggs can be candled to see if they are fertile. You should be able to see a little nucleus with red veins radiating outwards. If you have any clear eggs, throw them out now. Check again in another week, and again in week three. If you see a red ring or a big black spot it means that the eggs have stopped developing or have died in shell; get rid of those, too. (*See* section 'Progress inside Candled Eggs' above.)

Egg weighing is also a good way of tracking development, and I've done this for a hatch or two with goose eggs. Most bird eggs need to lose 13 to 15 per cent of their weight by hatching time, and weighing the eggs will help you appreciate if your humidity has been too high or too low during incubation. Weigh your eggs before incubation. On day fourteen weigh them again – you are aiming for a loss of 7 per cent in weight at this point. If the eggs are too heavy, reduce the humidity. If they are too light (which is unlikely unless you live in a desert region), increase the humidity by adding some water to the incubator.

Hatching

Hatching is trickier for ducklings than for chicks, and more so for goslings, so even for those experienced with poultry, waterfowl are more demanding. The crucial factor is attaining adequate humidity in the last few days. Be aware that different breeds have varying incubation requirements, so talk to breeders for breed-specific guidance. High humidity in the hatching stage is needed to prevent the membranes drying too fast and becoming tough and difficult to tear as the bird hatches. Two days before hatching is due, stop your automatic cradle or other turning system (if you have one), stop turning the eggs manually (if you don't), and remove any dividers.

Make sure the water channels are full, and you can also put in some narrow, tall jam jars full of water (tall so that newly hatched ducklings or goslings can't topple in and drown), as in some incubators the water channels don't provide anything like enough evaporation and therefore humidity for ducklings and goslings (though yours might). However, if your incubator is packed full of eggs you can't do this, so instead use thick blotting or kitchen paper, soaked, with one end in the channel to wick up the water, or include a soaked sponge (again if there is room). Don't forget to ensure adequate ventilation at this stage – check your incubator instructions – or you may get late death in shell. To add a bit of contradictory advice, I have hatched successfully in entirely dry incubators more than once.

Duckling hatching in the small incubator.

Ducklings hatching in the large incubator.

At the point when the eggs no longer need turning, just before they start to pip (usually day twenty-seven for ducks, twenty-nine for geese), I move them from their forced-air incubator to a still-air incubator, which I use solely as a hatcher as it can take a lot of water in the base. I found that the fan-driven incubators used for the main period of incubation can dry things out too much at hatching time for waterfowl.

I want to write this bold and large: **BE PATIENT**. Don't be too quick to chuck away eggs that haven't hatched after the majority of ducklings or goslings are under a heat lamp, safe in the brooder. Sometimes it's not death, but late hatching, which can be up to seventy-two hours after the first lot are out. Yes, some birds are out of their shell and bobbing about, fluffy and dry, in twelve hours, while some take much longer, up to forty-eight hours from first pip – patience is a necessity with this activity. If you are concerned that the membrane is drying out round a semi-hatched bird and it is sticking hard to the bird, do intervene or it will suffocate – the membrane dries out, particularly in a fan-driven incubator, and acts like deadly glue.

But apart from this, and removing any ducklings that are twelve to twenty-four hours old in batches, don't open the incubator unless absolutely necessary. If you do, you get rid of all that nice humidity you have worked so hard to create. Leave a minimum of six hours between openings, and twelve is better, and if you have opened it, add more water to channels and remove any empty shells at that point. As an example, in a recent batch I hatched eleven out of twelve, and every egg was fertile at seven and twenty-one days. Even so, there was a gap of four days between the first and the last hatching; that's nature (and external temperature changes) for you.

The necessity for patience extends to the question of helping birds to hatch. Unless, as above, the inner membrane of the egg is sticking like glue to the bird, then don't. It can take up to two days for a duckling or gosling to progress from first pip to hatching, during which time the yolk sac is being absorbed by the chick. If you try to pick at the shell you will make the bird bleed, which can be fatal. If a bird hasn't hatched it may well not be strong enough to survive. Once in a while birds try and pip out of the wrong, pointy end of the egg, not the wide end, and it's worth taking a risk to help these if they are to stand any chance of survival.

BROODING

From the incubator/hatcher put the goslings or ducklings in a brooder (home-made, posh shop-bought, or an adapted cage such as a hamster cage or a rabbit hutch) somewhere safe – away from cats, dogs, magpies and rats. I can't stress this enough. Years ago I lost eighteen week-old ducklings in one night from an outside shed that I thought was rat proof and wasn't. Rats can get through fairly narrow-gauge mesh, so you may have to adapt your brooder if it is not in a 100 per cent safe place. The home-made

Home-made brooder.

Bought brooder.

Non-slip mat.

brooder pictured has narrow-gauge weldmesh on the top and front, and the floor is made of aluminium sheeting. Because of the narrow-gauge mesh this is used for newly hatched birds.

The bought brooder shown has advantages and disadvantages: water and feed are kept out of the main body of the brooder, which helps keep things cleaner, but the gauge is wide, and although we've never had a rat attack inside it, it is used for the larger, less vulnerable birds just in case. Although galvanized it has become rusty after about five years of seasonal use.

The home-made brooder box – height 28cm (11in), width 55cm (22in), length 90cm (35in) – isn't tall enough once goslings are two weeks old. The bought brooder box – height 35cm (14in), width 55cm (22in), length 95cm (37in) – works well until goslings and ducklings are old enough to be put into cage runs on grass (*see* Chapter 4).

In the brooder you want to create a non-slip surface for the newly hatched birds to avoid them developing splayed or spraddled legs, and there are a number of options that provide suitable traction – for example, a single layer of not-too-thick towel, an old tea towel, thick textured paper, or my favourite non-slip mats as found in ironmongers, garden centres and pound shops, cut to size and topped with softwood shavings.

Ducklings and goslings should be under heat for two to three weeks depending on the weather; if the weather is warm you can get away without heat during the day towards the end of that period. Start the temperature at 32°C (a garden thermometer is good for checking this) and reduce by one degree a day by slightly raising the heat lamp. If the baby birds are clustering together under the lamp you know they are too cold, if they are spread to the furthest corners of the brooder they are too hot; if they are warm enough they'll be acting normally (feeding, drinking, moving about, interacting or snoozing).

Use a ceramic bulb, also known as a heat emitter, rather than an infrared bulb as a heat lamp, as there is less chance of splayed legs, and the birds enjoy a natural light pattern; it's dark at night rather than providing round-the-clock red glare. Ceramic is the more expensive choice, but it has a positive effect on the health of the ducklings and goslings. Heat emitters in bulb or trough format come in many different wattages, from 60 watt to 1 kilowatt, and many options between. Between 150 to 250 watts will be suitable for most domestic small-scale brooders, but you may want larger depending on how many birds you are raising, the size of the brooder, and how cold the space in which you are rearing.

Ceramic bulbs in heat-lamp holders.

Safety brooder.

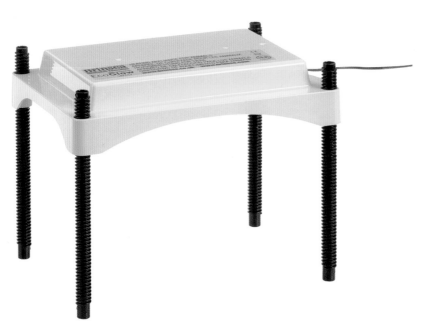

Goslings and ducklings.

Rearing a couple of ducklings in a very domestic environment.

Always make sure the brooder is large enough to allow the birds to get away from the heat source, and don't overcrowd it. An alternative is an 'electric hen' or safety brooder, raised to allow the birds to go underneath when they choose.

REARING

Newly hatched birds are still feeding off their internal yolk sac for the first twenty-four or more hours post hatching. Make sure that ducklings and goslings have access to drinking water in a purpose-made drinker, but don't give them a bowl of water to swim in until they start to feather up: they can easily drown as they don't produce their water-repelling oil or weather-proof feathers until they are some weeks older. Waterfowl raised naturally take on oils from their mother as she sits and grooms them.

Feed ducklings and goslings ad-lib on non-medicated chick crumbs or waterfowl crumbs if you can get them for the first three weeks (an eggbox makes a cheap, biodegradable, throwaway food container), then move on to grower pellets (*see* Chapter 8 for details on feeding). At this age make sure their food is close to their drinker. It is a very messy business and the brooder will need daily cleaning out, but the youngsters need to have water to help the food go down; they tend to shovel some food, take a drink, shovel food, drink again.

Make sure their drinker is refreshed regularly at least twice a day, three times if possible, and isn't clogged up with feed or shavings, preventing access to the water. Goslings should also have access to greens: some freshly cut grass clippings, chopped spinach or lettuce and dandelion leaves will be welcome.

At four weeks old, now off heat, they can go outside in a predator-proof pen and hut. Once they are eight weeks old they need deeper water to dip their heads into (shallow buckets are good, the rubber type for feeding horses), and preferably enough to swim in.

CHAPTER THIRTEEN

Making the Most of Your Ducks and Geese

You could justifiably say that the pleasure of owning, rearing, admiring and knowing your birds is more than enough reward for the care you give them. Whether it's holding a day-old duckling in the palm of your hand, its brand new, soft, unweathered webbed feet tickling your skin, or appreciating the expected warning honk of your oldest gander as you lift an egg from his mate's nest, every day can bring a smile with ducks and geese.

However, there are additional benefits that might appeal beyond the enjoyment of the birds and eating their meat and eggs. There are opportunities for the inveterate salesperson through the simple selling of eggs and birds, and for the crafter, whether that be the sumptuous decoration of the eggs, or using feathers in collages or wall hangings and in a host of personal decorative purposes, from jewellery to hats and boas.

USES FOR FEATHERS AND DOWN

The soft downy underlayer of geese and ducks makes for the most luxurious pillows, quilts, duvets, eiderdowns, comforters, cushions and bolsters. Eiderdowns are specifically made from the down of the Eider duck, taken from the nest once the ducklings have hatched and dispersed. Just four tons of eiderdown is produced globally each year, with 75 per cent of that coming

Swimbridge, August 1985. Documentary photo: James Ravilious/Beaford Archive

from Iceland, and the awe-inspiring price of the resulting pillows and duvets (we're talking thousands of pounds) reflects the scarcity of the filling.

In a time of hi-tech cold weather clothing, the best quality jackets and coats are still filled with

OPPOSITE: *Carved egg.*

Sitting with your geese.

Luxury Eider pillow.

Uses for Feathers

- Arrow flights
- Shuttlecocks
- Angel wings or head-dresses for school plays
- Pen quills
- Christmas garlands, wreaths and tree decorations
- Necklaces (perhaps rather tickly), brooches and earrings
- Mobiles
- Trimmings for dresses, hats and fascinators
- Hair-extension decorations
- Present-wrapping decorations
- Bookmarks
- Stuffing for draught excluders (for cushions, pillows, bolsters and duvets it would be a lot more comfortable using the down)
- Feather dusters
- Fans

Gathering Eider down in Iceland.

Duck-feather bow tie.

goose down to give a very warm but lightweight garment. Duck down is also used and is the cheaper option. The down jacket and gilet are as much a fashion item as a utility one, with many more worn by the urban trend-setters and cosy conscious than by tough explorers in the Arctic. Down-filled sleeping bags are also costly items, lightweight and able to be packed compactly, which is convenient if you hope to sleep in comfort on the side of a mountain. The truly dextrous waterfowl keeper could make a fortune constructing their own down-filled items.

Feathers should be washed and dried (tightly tied in an old pillowcase to avoid a permanently fluffy washing machine) before use, and sterilized to kill off any risk of viruses, mites or bacteria. Clean feathers can be baked in the oven at 120°C for 2.5 hours, or boiled for twenty minutes. Feathers should never be plucked from live birds; make use of the ones they naturally preen and leave on the ground, and the plentiful supply that results when you pluck birds that have been slaughtered for meat.

DECORATED EGGS

Fabergé eggs are world-renowned jewelled eggs created by the goldsmith Peter Carl Fabergé between 1885 and 1917, made for the Russian tsars Alexander III and Nicholas II as Easter gifts for their wives and mother. The Fabergé jewellery company continues to make jewelled egg-shaped charms and pendants, while artists and crafters across the globe take inspiration from the original eggs, working with duck and goose eggs to create works of art that are no bigger than your hand. Eggs are painted, decorated and carved. The carved or painted egg used as part of Easter celebrations is known as a 'pysanka' in the Ukraine. There is even an egg-shaped Pysanka Museum in the western Ukrainian city of Kolomyia, dedicated to a collection of over 10,000 pysanky.

HATCHING EGGS SALES

There is an apparently unending demand for hatching eggs: that is, eggs that are likely to be

Fabergé Renaissance egg.

Carved goose egg.

Easter egg with a strawberry pattern. Goose egg, wax-resist dyeing. Ukraine 2020.

The Pysanka Museum in Kolomyia, Ukraine.

fertile, which can be hatched naturally under a broody, or in an incubator. If you have quality birds that produce more eggs than you can use, selling the surplus can, at the very least, make a significant contribution towards your feed costs. Local customers will ask to see your birds when collecting eggs, and for those purchasing from further afield ensure that you package eggs appropriately to minimize the ever-present risk of damage in transit.

Make sure that the eggs you sell are fresh, so the buyer will receive eggs that are no more than a week old, and ensure that the eggs are definitely from the breed of birds being advertised. Also, do check the fertility of your eggs at various points during the laying season so that you can be confident in your product. You cannot, and should not, guarantee 100 per cent hatchability as there are so many variations in people's aptitude and expectations for hatching, and nature will always have the last laugh.

BIRD SALES

Plenty of customers prefer to buy young birds, and leave the demands of hatching to the breeder. Some will want very young birds from a day to a week old, others will be after pairs (male and female), or trios (one male and two females) that are just entering their first breeding season. Some purchasers will be very experienced, others just starting out, and you need to be prepared for a myriad of follow-up questions from the latter as they find their way.

USE THAT FAT!

'Schmaltz' is a word I knew long before I'd heard the term goose (or chicken or duck) fat. It conjured up deliciousness, the scrapings from the roasting pan that were spread on toast, went into liver pâtés, created the crispest roast potatoes, or added to a humble dish of poultry leftovers tossed with rice and a generous sprinkling of vegetables. Once you've enjoyed

Goose fat.

your roast bird, drain the resulting duck or goose fat from the roasting pan into a wide-mouthed jug, and then sieve or filter it through muslin and pour it into a sterilized jam jar; it will keep in the fridge for many months, or you can freeze it for a great deal longer.

There are plenty of non-culinary uses for fat, and you may not feel the need for some, or any of these; several of the uses below should come with a health warning! Open water or cross-channel swimmers are renowned not just for their perseverance and endurance, but for liberally covering themselves in goose fat to protect them from the cold and to avoid chafing. The old cold and decongestion remedy of rubbing goose grease on your chest just makes me think of sticky pyjamas, but it has been used for centuries to relieve minor aches and pains.

I also remember being asked by a farmer friend to save him the goose grease from our Christmas dinner: he wanted it to soothe the hardworking, tired legs of his rather fabulous horses. Used for any swellings that were caused by a knock, the goose grease was applied to reduce the swelling in a haematoma, so that the horse wasn't left with a hard lump. Rubbed in gently – not too vigorously or it blisters – leave for five days, and only repeat if necessary.

There are even suggestions that it is a natural anti-depressant, although how fat is used for that effect I can't surmise. How to apply it as treatment for piles is a little clearer! Goose and duck fat can be more practically made into soaps and creams to treat skin conditions.

Goose fat soap.

CHAPTER FOURTEEN

Recipes

Growing up, duck was what my parents had on the very rare occasions we were lucky enough to go to a restaurant; it was always roasted and covered with orange sauce (which, not being a fan of marmalade flavours, I disliked intensely). And then cherry sauce became an option, and my world view of duck changed to one of delight and anticipation; I am nothing if not a child of the sixties and seventies.

The more frequently enjoyed duck treat was that offered by the Chinese takeaways in North London (and everywhere else), cooked to tenderness with crisp skin, coated with a smear of plum sauce, accompanied by thin strips of cucumber and spring onion if you were so inclined (I was not), wrapped in paper-thin pancakes, and scoffed with huge grins. I've always liked eating with my hands, and Peking

OPPOSITE: *Duck breast cooked to perfection.*

ABOVE: *Cooked duck leg.*

duck offers everything I love about food: a bit of at-the-table prep, a sharing platter, followed by all that crunch, spicy sauce, sweetness, savoury loveliness, stickiness and informality. Duck was the food of memories, for special occasions and treats out, and always cooked by someone else.

We had goose once, as an alternative to turkey one Christmas, a never-to-be repeated adventure because of the smoked-out kitchen, as no one had thought to pour off the excess goose fat as it accumulated in the roasting pan. A couple of times my father brought home goose eggs to make himself the most gigantic dish of scrambled eggs, but I don't recall eating it, or being offered a duck egg as a more modest option.

Decades on, duck and goose are still a treat, and the whole birds are inevitably roasted, because that's what we love, in particular the contrast of crispy skin and succulent flesh, and melting the fat off the bird in a way that does the flesh justice. But there are many ways to ring the changes, even if that's your preference for serving waterfowl. And there's no need to create a pall of smoke when cooking these wonderful birds.

Below are a few recipes for cooking whole birds, and also for legs and breasts, which make delicious quick meals, plus a few ideas for using other parts of the bird that may be less familiar. The dishes are a mix of old-fashioned favourites and more modern approaches, and I hope you enjoy making them.

PEKING DUCK

There are cheating ways to do this – simply rubbing the skin with soy sauce and honey, and then roasting – but you won't get the texture you really crave from Peking duck, so the following recipe is worth the extra time and effort. You need to start this dish the day before

Peking duck.

you want to eat it, which is always something of a challenge. This recipe serves four to six people, depending on whether you see it as a starter or the main event, and of course on the size of the duck.

Ingredients:
An oven-ready meat duck
1ltr water
150ml balsamic vinegar
3 tablespoons runny honey
3 tablespoons dark soy sauce
1 tablespoon sea salt
2 tablespoons Chinese five spice
2 tablespoons ground Sichuan pepper

For serving: Chinese pancakes (you can buy these in any Chinese supermarket), spring onions, cucumber, hoisin or plum or sweet bean sauce (or all three).

Method:
• Make sure your duck is completely dry, and if not, pat away any moisture with kitchen paper.
• Bring the water and vinegar to the boil in a large saucepan, and either hold the duck over the pan and ladle the mixture over it until it is completely coated, or dip it in the pot several times to coat it. Keep the mixture aside while you put the duck in a cool, well-ventilated place to dry for several hours, or preferably overnight.
• Next day, bring the reserved water-vinegar liquid to the boil, add the honey and soy sauce, and again bathe the duck skin; leave to dry for a further two to three hours.
• Mix the salt, five-spice and Sichuan pepper, and rub it inside the cavity.
• Put the duck on a rack in a roasting tray, and pour 150ml of water into the pan. Roast for 20–30 minutes in a hot oven at 230°C/450°F/gas 8, then lower the oven temperature to 180°C/350°F/gas 4, and roast for another hour.

• When cooked (so that clear juices spurt out when you pierce a skewer through the leg), take it out of the oven and rest for ten minutes.
• Carve on to a hot plate, and serve with the warm pancakes, sauce(s), and onions and cucumber. You'll need fingerbowls and napkins!

ROAST DUCK WITH CHERRY SAUCE

The duck can be stuffed or not for this recipe, but just as with roast goose, stuff the neck end, not the cavity. I'd choose a simple stuffing of breadcrumbs, chopped sweated onion, the chopped duck liver, a couple of spoonfuls of Morello cherry jam, fresh thyme, salt and pepper, mixed well.

For the cherry sauce:
2 tins of black pitted cherries in syrup
2 tablespoons Morello cherry jam
1 teaspoon balsamic vinegar
2 tablespoons cornflour
2 or 3 shallots
Giblet stock made from the neck, gizzard and heart

Method:
• Prick the duck skin all over with a sharp fork, and rub with sea salt.
• Roast the duck in a hot oven, breast side down on a rack sitting in a roasting tin for half an hour, then turn it over and continue to cook, pouring off any fat that collects in the tin every half hour. When cooked the legs should pull away easily from the body.
• Drain off any remaining fat, turn up the heat in the oven, and cook the duck for 15 minutes more to get the skin nice and crispy. Rest for 15 minutes before carving, which gives you time to finish the sauce.

Sauce:
• Heat half a tablespoon of the duck fat in a

saucepan, and cook the finely chopped shallots until lightly golden.

- Add the juice from the tinned cherries and giblet stock (a total of 750ml of liquid), and a couple of tablespoons of Morello cherry jam and a teaspoon of balsamic vinegar; stir and cook until the liquid reduces.
- Mix a couple of tablespoons of cornflour with a little water, then pour in a bit of the gravy, stir, and then put the cornflour mix in with the gravy, stirring all the while. Add the drained, pitted black cherries and carefully bring to a simmer.
- Season as needed, and reduce until the sauce coats the back of a spoon.
- Serve in a blaze of glory wearing your best 1970s kaftan.

THE PERFECT PINK DUCK BREAST

I don't know about you, but although I like a touch of pink to my duck breast, I don't want it bloody or half raw.

Method:
- Cook in a pan or griddle that is already properly hot before you put in the meat, and where the dab of oil you're cooking it in is also hot (but not smoking). If the breasts are chunky or someone prefers no pink, spatchcock each one (split almost in half, with a hinge of meat holding it together), either before or halfway through cooking.
- Put in the pan skin side down, and don't overfill the pan. Let the skin get nice and crisp before you turn them.
- Timings depend on the size of the breast, the heat, and personal preference, but in every case let the breasts rest for five minutes before serving.
- For something really juicy, add chunks of fresh pineapple, some runny honey, and a generous dash of dark soy sauce into the pan juices; stir gently, picking up all the meaty bits, and pour the lot over the breasts to serve.

CONFIT (PRESERVED) DUCK OR GOOSE LEGS

Ingredients:
2 goose or duck legs
Sea salt
Pepper
Handful of fresh thyme
3 or 4 cloves of garlic sliced (if you desire)
2 bay leaves
750g goose or duck fat
2 tablespoons of olive oil

You will need a Kilner jar large enough to contain the legs. If you've cooked the breasts as a separate meal, you can preserve the legs in the reserved duck or goose fat.

Method:
- In a bowl, rub the salt, pepper, thyme, bay and garlic into the legs, and put in the fridge for forty-eight hours; give the legs another massage with the mix halfway through.
- Heat a couple of tablespoons of olive oil in a cast-iron casserole, and brown the legs all over. Add in the seasoning mix, cover the legs with the duck/goose fat, and put the casserole, covered, into a medium hot oven for two hours.
- Leave to cool a little, then when still warm put the legs and the seasoned/fat mix into a Kilner jar. When cold, put in the fridge or a cool pantry, where the legs will keep for two months or longer.
- The legs should be eaten hot, brought to temperature in a little of the fat, and served with lentils or mash.
- The seasoned fat mixture can be used again, preferably reheated and the bits filtered out first. The preserved legs are a necessary ingredient for cassoulet.

CASSOULET

The first time I had cassoulet was from a huge tin bought in a French supermarket. It was a fairly upmarket affair as tins go, and eaten bubblingly hot with relish sitting outside in an isolated campsite – but making your own

Cassoulet.

is rather wonderful, as it's a bit of a meaty process, literally and metaphorically. It's not what you'd call a light dish: to be honest it's as gut-busting as they come, and to be reserved for cold winter evenings when you've been working outside all day and in need of a good feed. There are many recipe variations, but goose or duck, pork and beans, sausage and tomatoes are staples. The quantity below serves four to six people, and is adapted from Jane Grigson's book on *Charcuterie and French Pork Cookery*.

Ingredients:

For cooking the beans:
450g dried haricot beans
250g pork or bacon skin cut into small pieces
Knuckle of pork
450g pork belly
Large onion (peeled and stuck with four cloves)
Large carrot
Bouquet garni (bay, thyme and parsley stalks)
4 garlic cloves
450g of Toulouse sausages

For cooking the pork and goose (or duck):
450g shoulder of pork
750–900g preserved duck and/or goose legs
3 large onions, chopped
Garlic, a few or many cloves, to your taste
Bouquet garni
Black pepper
4 large or 8 medium tomatoes cut into chunks
3 tablespoons tomato purée
Beef stock
Goose or duck fat (from the preserved leg jar)
White breadcrumbs

Method:
• Soak the beans in water overnight. Drain and put in a casserole (traditionally an earthenware pot) with the pork skin, onion, carrot, pork knuckle bone, pork belly, herbs and seasoning, and cover with water. Bring to the boil and cook in the oven until the beans are tender (about 1.5 hours), adding the sausages for the last half hour.
• While that's cooking put the onions in a large frying pan with some duck/goose fat until they are golden. Cut the goose/duck and pork into mouthful-sized chunks, and add to the onion to brown. Pour off any surplus fat, and add the tomatoes and a bit of the stock, the tomato purée and seasonings/herbs. Simmer, cover and cook for half an hour.
• Drain the beans when cooked (keep the liquid), take out the bouquet garni and knuckle bone, and cut the pork belly into chunks. Traditionally the pork or bacon skins are then used to line the pot for the final cooking.
• Put half the beans into the pot, then the pork and goose/duck mix, the meat from the bean mix and the sausages, and then top with the rest of the beans. Pour over a quarter of the bean liquid, and top with a centimetre layer of breadcrumbs, dotted with bits of duck or goose fat.
• Cook slowly in a low oven (about 160°C/320°F/gas mark 3) for 1.5 hours until the crust is golden. You can push the crumb layer down with a spoon while cooking and top up with more breadcrumbs, and add more bean liquid if it gets too dry, but this is not meant to be a wet stew, so take it easy.

No accompaniment is needed. And it's unlikely you'll have room for pudding.

ROAST GOOSE

I am not going to tamper in any way with tradition here: roast goose is simple, it's glorious, and you don't really need a recipe. Your fellow diners will thank you.

Serve with red cabbage casseroled with apples, red wine and sultanas, baked or mashed potatoes (keep the creamy dauphinoise or goose fat roasties for the following day, served with the goose leftovers), leeks braised with peas, a

*Roast goose –
halfway there.*

simple dollop of crab-apple jelly and a generous bowlful of bread sauce.

I'd advise against stuffing the cavity as it would be heavily imbued with goose fat, and you can have too much of a good thing, but if roasted on a trivet (you really should use one for roasting goose or duck), then put stuffing in the flap of skin at the neck end as you can still get a good quantity in there.

STUFFED GOOSE NECK

I'm very conscious that just the name of this dish will deter a lot of people, but stuffed duck or goose neck was a childhood highlight for me, and a necessary accompaniment to a posh roast bird dinner, so perfect for Christmas (even though this is fundamentally a Jewish dish). The critical ingredient is the tubular skin of the bird's neck, impossible to get from any supermarket, but easy to get hold of if you slaughter your own birds, as we do, or buy birds long-legged (plucked but not dressed, with head and feet intact).

When butchering the bird, pluck well down the neck to keep it as long as possible, remove the head, make your cut around the neck close to the body, and then roll the skin off the neck, keeping the skin as a tube. Basically, this dish is a bolster-shaped dumpling, roasted in the bird's neck skin – what's not to like?

For the stuffing:
4oz breadcrumbs or flour, or a mix of both
2oz suet or duck/goose fat
2–4 teaspoons of fresh herbs to your taste (I like thyme and sage)
Half a lemon, zest and juice
1 egg
1 small onion chopped finely
Thinly chopped liver from the bird whose neck you're stuffing
Salt and pepper

Method:
• Mix all the ingredients together, adding a little water to moisten if necessary to hold the mix together.

Stuffed goose neck before cooking.

- Sew up one end of the goose neck with butcher's needle and string, fill with the stuffing, and sew up the other end securely.
- Put in the oven in the roasting tin with the goose, about an hour before you are ready to take the bird out of the oven, basting occasionally. If there's no room, put it in its own dish and bake in a moderate oven until crisp and brown.
- Serve in slices as an accompaniment to roast goose.

The neck skin can also be used as a casing for goose salami, a typical Italian product; and in the spirit of goose charcuterie imitating pork, the skin can be cooked in goose fat until crisp and sprinkled with salt to serve as crackling, while the breast can be cured in its skin like a ham.

WARM DUCK OR GOOSE SALADS

It might be passé, but I continue to be a fan of warm meat salads, and leftover duck or goose shreds of meat and crispy skin are perfect main features for a good bowlful of crunch and texture.

Any mix of salad leaves will do, with artichoke hearts or broad beans a lovely addition. Most important is the dressing – a sweet rather than a sour one. Because we make our own cider, we also have cider vinegar, which I transform into blackberry or raspberry vinegar.

Method:
Put two parts olive oil to one part fruit vinegar in the bottom of the salad bowl. Toss the vegetable salad ingredients in it, quickly heat pieces of cooked meat and skin in a frying pan, and when hot, strew them into the bowl, toss, and eat!

EGG RECIPES

You can make absolutely any egg dish with duck or goose eggs; they definitely don't need to be baked and disguised with sugar in a sponge or tart, although they are good for those, too. Use them in any of your favourite egg recipes: quiches, cakes, scrambled with smoked salmon, pancakes, eggs and bacon, custard, egg pasta, fish pie, kedgeree, scotch eggs, frittata, crème brûlée, French toast, Shakshuka, meringue,

Duck eggs.

Fried duck eggs.

lemon curd, and so much more. Goose eggs are particularly favoured for making fresh egg pasta.

Egg Sizes

There is significant variation depending on the age of the bird (eggs get larger as a bird ages) and the size of the breed, but you can expect a goose egg to weigh 150–200g and a duck egg to weigh 70–100g, in comparison to a medium hen's egg weighing 53–63g or a large one at 63–73g. Any double-yolkers will be significantly larger and heavier, our Aylesbury ducks prolifically producing double-yolked eggs as large as a goose egg, weighing up to 160g.

Adjust recipes in your favourite cookery books to take these differences into account, but unless it's a delicately balanced recipe such as pastry and some breads and cakes, using the greater volume of a duck egg instead of a hen's egg won't matter too much. In recipes asking for six or more eggs, you might want to reduce by one – but to be honest, in most cases, the more egg the better!

DUCK EGG MAYONNAISE

This is one of my favourite ways of eating duck eggs (other than eaten boiled and hot with buttered sourdough toast).

Method:

• Boil the eggs for ten minutes if large breed (nine if smaller). Cool them in cold water.
• Get some cucumbers and peel off and discard the skin with a potato peeler, then peel the cucumber flesh into ribbons straight on to a big plate.
• Peel and halve the eggs lengthways. Spoon over mayonnaise (*see* below), thinned with a little water or freshly squeezed lime or lemon juice so the mayo coats the eggs, rather than sitting on top like a party hat.
• Sprinkle chopped chives over, and eat with a wedge of good bread.

This looks amazingly indulgent if you make a towering platterful for guests: when you have more eggs than you know what to do with, it's time to have a party!

Mayonnaise ingredients:

Duck egg yolks (two if from a large breed, three from smaller breeds)
300ml groundnut oil
Lemon or lime juice (barely a teaspoon)

Method:

Put the yolks in a bowl with a pinch of salt and pepper, and whisk in the oil, a drip at a time to start with, turning to a trickle as the mix thickens. When the oil is incorporated, add a little lemon juice and stir in.

You could do half groundnut oil (fairly neutral) and half olive oil, which will give more punch.

DUCK EGG BRIOCHE OR CHALLAH

Heavily enriched, buttery bread: a serious teatime or breakfast treat.

375g strong white flour
2.5 teaspoons dried yeast
2 tablespoons sugar
1 teaspoon salt
6 duck eggs (or 3 goose eggs)
200g softened unsalted butter (reserve a little to grease the tin)

• Mix the yeast with a couple of tablespoons of water and put to one side. Put the flour, sugar and salt into a large mixing bowl. Make a well in the centre of the flour mix, and add the yeasty water and five of the duck eggs (beaten); mix together to make a soft, damp dough. Turn the dough on to a floured surface and knead for ten minutes until it is nice and elastic.
• Wash and dry your mixing bowl, then grease

Challah.

it with a knob of melted butter (taken from the 200g). Put your dough in the bowl and turn it to make it nice and buttery all over. Cover the bowl with a clean tea towel, and leave the dough to rise for an hour or more in a warm place to double in size. Then using your knuckles, knock it back and leave to rest for another ten minutes.

• By hand, press small nuts of butter into the dough, until you've added 175g. Turn out again on to a floured surface and knead for a few minutes until the butter is evenly incorporated.

• Either split the dough into three pieces, roll out each piece and plait into a loaf for a challah loaf and put on to a baking sheet, or divide the dough into ten balls and place them into a buttered loaf tin. Cover with your tea towel and let the dough double in size once more; this should take half an hour, or more depending on the ambient temperature.

• Heat the oven to 220°C/425°F/gas7. Brush the loaf with an egg yolk mixed with a dash of water and bake for 20–25 minutes. Leave to cool on a wire rack, or, like us, tear off hunks of brioche and eat warm with raspberry jam.

PICKLED EGGS

Pickled eggs are ubiquitous behind the counter of most fish-and-chip shops, and in many pubs will be made with chicken eggs, but there's absolutely no reason why you can't use duck (or goose) eggs for this somewhat acquired taste.

Ingredients:
12 shelled, hard-boiled duck eggs (8min if you like the yolks soft, 10min if you don't)
4 garlic cloves
Thinly sliced onion
Bayleaf
Handful of your favourite fresh herbs, or you

can add spices (chilli flakes, turmeric, curry powder, coriander or mustard seeds)
1 teaspoon whole black peppercorns
1 tablespoon sea salt
100g caster sugar (you can omit if you prefer)
400ml apple cider vinegar (you can use malt vinegar if you prefer)

Method:
- Layer your shelled eggs into a sterilized preserving jar with the garlic and herbs.
- Make the brine with 500ml water in a pan plus the vinegar, peppercorns, sugar and salt, and bring to the boil.
- Once the brine is cold, pour it into the jar to cover the eggs, tapping out all the air bubbles so they rise to the surface.
- Put on a lid and store in a cool, dark place for at least two weeks, for a maximum of three months.
- Once the jar is opened, keep in the fridge and eat within two weeks.

OFFAL

Livers
Duck and goose livers can be used in any recipe for chicken livers, whether that be chopped liver, pâté, mousse, parfait, or served fried with shallots and marsala or madeira on chunky sourdough as a starter or lunch, or on tagliatelle as a main course. Or how about using them as a stuffing for a great fat raviolo or two? The only issue is having enough of these delicious livers to make a meal. The liver is also wonderful chopped and added to any stuffing.

Chopped Liver (Duck or Goose)
More normally made in Jewish homes with chicken livers, this wonderful easy supper dish is perfect when adapted for goose or duck liver.

Ingredients:
1 or 2 duck or goose livers
1 or 2 hard-boiled duck eggs, chopped
1 medium onion, chopped
Duck or goose fat

Method:
- Fry the onion in plenty of duck or goose fat until it softens, and then add the livers. Don't overcook – they should be just pink inside, but not bloody.
- Leave to cool a little in the pan, then tip on to a board and chop.
- Mix with the hard-boiled eggs until you have a coarse, pâté-like texture. Every little bit should be scraped carefully into a bowl, including any fat still lingering in the pan.

Eat this slightly warm with toast. Heaven.

Hearts and Gizzards
I've always had a love of poultry hearts, and they rarely get the chance to make it to the table.

Method:
- Add to the giblet stock to make great gravy, or roast with the bird. Either way, eat as you serve the meal: it's the cook's treat for all the effort made. If cooked long enough in the stock and the bird is young, the gizzard will be tender.

It's always a toss-up if the dogs get a bit or not, as it might go the way of the heart.

Tongues
Having our own cattle, ox tongue is something of a treat relished every few months. As a child my mother would cook a lovely dish of lambs' tongues, but the restaurant and export trade seem to have removed this option from the butcher's counter. Duck and goose tongues may not be something you see on Western plates (the West is undoubtedly overly squeamish about offal), but they are certainly found in Chinese recipes. To make them worth the eating you

Duck tongues.

Boiling the duck tongues.

need a fair number, so store them in the freezer until you have twenty or so.

Method:

- As for any tongue, duck and goose tongues need to be simmered for a long time, so put in a pot of water or stock, bring to a simmer, then put in the oven for several hours until tender when pierced with the tip of a sharp knife.
- When cooked and still hot, remove the bone at the root of the tongue. Make a spicey teriyaki or hoisin or plum sauce, or a mix of shallot, garlic and honey, and fry the tongues in this; add in some chopped pak choi and Chinese mushrooms, and serve on a bed of rice.
- Alternatively, make a marinaded cold tongue dish by boiling the cooked tongues in water for five minutes, then pour away the water and add a bottle of Chinese marinade sauce. Heat the sauce and tongues to boiling point, then turn off the heat and leave the tongues

to marinade for at least two hours before serving.

Feet

Feet are even less likely to be consumed than tongues. My mother used to enjoy chicken's feet cooked long and slow in broth, but the rest of the family didn't fight over them. Again, the Chinese are ahead of the game, seeing feet of all kinds as a delicacy and not to be wasted. Cooking feet is a fairly long process, starting with washing them thoroughly and clipping off the toe nails. Two recipes are given below.

Method 1:

- Once cleaned, prepared and dried, dip in sugar or flour, then deep fry for ten minutes, before plunging into cold water to puff up.
- Simmer the feet in a pot of stock with added root ginger for a couple of hours.
- Drain, and put the feet into a spicey marinade (soy sauce, oyster sauce, garlic, chilli flakes,

Marinaded duck tongues
ready to serve.

Duck feet.

Part-cooked duck feet.

Cooking the duck feet until tender.

Stewed duck feet and shiitake mushrooms in abalone sauce.

rice wine and a bit of sugar) in the fridge overnight.
• Finally, heat the feet and marinade in a wok, and eat with sticky rice, with a fingerbowl and napkin to hand.

Method 2:
The feet of ducks and geese can also be braised long and slow, producing a tender yet chewy texture.

• Put the feet in a heat-proof bowl over a pan of hot water with fresh ginger, star anise, spring onions and water, and steam for 1.5 hours.
• Fry more ginger, garlic and chopped shallots, add the steamed feet and cook until browned.
• Then add rice wine, Chinese mushrooms and oyster sauce, and cook until sticky.

Glossary

Abscess – localized collection of pus.

Acre – imperial unit of area where one acre equals 4,840 square yards (4,047 square metres) or 0.405 hectare.

Anaemia – lower than normal number of red blood cells.

Anatidae – the family of birds including ducks, geese and swans.

Angel (or slipped) wing – single or both wings hanging down, caused by too much protein in feed.

Anserine – goose-like, relating to or resembling a goose.

Anseriformes – the order of birds including Anatidae and others.

Anthelmintic – drug which kills certain types of intestinal worms, also known as wormer.

Antibiotic – drug which inhibits the growth of/destroys micro-organisms/bacteria.

Antibodies – proteins produced by the immune system to fight specific bacteria, viruses, or other antigens.

Antitoxin – antibody that can neutralize a specific toxin.

Aspergillosis – respiratory disease.

Autosexing – the colour of the bird reflects its gender.

Avian Tuberculosis – rare disease that can be transmitted by wild birds.

Bean – hard, raised bump at the end of the bill.

Body Condition Score (BCS) – numeric value given to the degree of fatness/thinness and condition of an animal's body.

Botulism – disease transmitted from access to toxins.

Breed – group of animals with similar characteristics distinguishing it from other animals, that are passed from the parents to the offspring.

Brood – the young hatched from a single clutch.

Broody – female bird who ceases to lay and shows willingness to sit on eggs and rear offspring.

Buffled – fluffy feathers on the head of call ducks.

Bumblefoot – hard swelling on the underside of the foot.

Candling – illuminating an egg to see contents.

Carcass/carcase – dressed body of a slaughtered animal (intestines, head, feet and skin removed).

Caruncles – fleshy parts on the head or neck (for example on the Muscovy duck).

Cellulose – component of plant cell walls, which is not digestible by most animals.

Cere – fleshy growth of the bill above the nostrils.

Cloaca – cavity just inside the vent in which the urinary tract, reproductive tract and digestive tracts meet.

Closed flock – no new birds are introduced into the flock.

Clostridial diseases – potentially fatal

infections caused by clostridia (soil borne) bacteria.

Clutch – number of eggs laid by a female for incubating.

Coccidiostat – any of a group of chemical agents mixed in feed or drinking water to control coccidiosis in animals.

Concentrate – high energy, low fibre feed that is highly digestible.

Conformation – combination of structural correctness and muscling of the animal including its frame and shape.

CPH number – County Parish Holding number, which registers land as being for agricultural use.

Creep feed – small pellets of high-protein supplementary feed given to young animals.

Cross-breed – animal whose parents are of two different breeds.

Cross-grazing – using two or more species of animals on the same land, because they graze in different ways and benefit the sward.

Cull – animal no longer suitable for breeding, which is used for meat or disposed of.

Culling – slaughtering an unwanted animal.

Dewlap – flap of skin that hangs beneath the lower jaw or neck.

Diarrhoea – also known as scouring; an unusually loose or fast faecal excretion.

Down – coat of newly hatched ducklings and goslings and the undercoat of adults.

Drake – male duck.

Drench – orally-administered liquid medicine (n); to administer a liquid medicine (v).

Dressing percentage (also killing out percentage) – the percentage of the live animal that ends up as carcass.

Dropped tongue – condition where the lower part of the mouth is engorged with food waste due to inadequate water supply.

Dropped willy (penis paralysis) – unretracted, protruding penis in drakes and ganders.

Droving/driving – walking livestock including geese from one location to another.

Duck – generic term for a member of the Anatidae family. Can be used more specifically to refer to females.

Duckling – young duck up to the age that it gets its first proper feathers.

Duck Virus Enteritis (DVE) or duck plague – serious disease transmitted by wild mallard to domestic waterfowl.

Dump nesting – sharing of a nest by multiple females.

E-Coli – disease spread by infected droppings and dirty eggshells.

Eclipse moult – moult at the end of the summer. For males this takes them from smart colourful feathers to drab camouflaged ones.

Egg – organic vessel laid by female birds containing the germ of a new individual, enclosed within a shell.

Egg tooth – horny end of the bill that enables a duckling or gosling to hatch out of the egg. It falls off within a few days of hatching.

Embryo – an animal in the early stage of development.

Eviscerating – removing the intestines in preparation for cooking and eating.

External parasites – fleas, lice, mites and ticks that feed on body tissue such as blood, skin and feather resulting in irritation, blood loss and disease.

Faecal egg count (FEC) – the process of assessing the level of parasite load based on the number and type of parasite eggs found in the faeces.

Faeces – manure or excrement produced by an animal.

Fertilizer – natural or synthetic soil improvers which are spread on pasture to improve fertility.

Finish (condition) – amount of external fat that covers the body.

Fledging period – the time between hatching and first flight.

Flightless period – time between moulting and regrowth of flight feathers.

Flights – the large wing feathers including primaries and secondaries.

Flock – a group of birds (or sheep).

Fodder crop – a plant grown for animal feed.

Food chain information – regulation requirements if you are intending to slaughter animals for human consumption.

Forage – edible plant material used as livestock feed.

Fowl or Duck Cholera (*Pasteurellosis*) – respiratory infection spread by rodents and wild birds.

Free range – birds raised in conditions of freedom as to foraging, rest, bathing, feeding and nesting.

Gaggle – a group of geese.

Gander – male goose.

Goose – female goose (or generic term for the species).

Goose Parvovirus (**GPV**) – disease with up to 100% mortality rate.

Gosling – young goose.

Halal – set of Islamic dietary laws that regulate the preparation of food.

Heat stress (**staggers, sun stroke**) – overheated birds, needing quick treatment to avoid fatality.

Hectare – metric unit of area equal to 10,000 square metres, or 2.471 acres.

Hen – female duck.

Heritability – extent to which a trait is influenced by genetic make-up.

Hot carcass weight – weight of a dressed (eviscerated) carcass immediately after slaughter prior to the shrinkage that occurs in the chiller.

Hybrid vigour – increase in performance due to cross-breeding.

Hypothermia – condition characterized by low body temperature.

Immunity – natural or acquired resistance to specific diseases.

Incubation – applying heat to an egg to enable it to develop and hatch.

Incubation period – number of days for eggs to hatch once heated to the correct temperature.

Incubator – artificial means of incubating and hatching eggs.

Inbreeding – mating or crossing of closely-related animals; sometimes referred to as line-breeding when carried out to pass on or strengthen certain desirable traits.

Internal parasites – parasites located in the gastrointestinal system in animals.

Intramuscular (**IM**) **injection** – one which is given straight into a muscle.

Intravenous (**IV**) **injection** – one which is given directly into a vein.

Jugular – veins in the neck, which carry deoxygenated blood from the head back to the heart.

Keel – on birds the equivalent of the brisket, above the breast. Particularly prominent in Aylesbury ducks and Exhibition Toulouse geese.

Knob – prominent bump, swelling, bulge, or projection at the base of the bill in some breeds such as the African and Chinese goose, typically positioned on the upper mandible.

Kosher – food prepared in accordance with Jewish dietary laws.

Lameness – often the first sign of a waterfowl disease, or a mechanical injury.

Line-breeding (*see also* 'inbreeding') – mating of closely-related animals within a particular line; inbreeding.

Mandible – upper (maxilla) and lower (mandible) makes up the top and bottom part of the bird's bill.

Masculinity – describes the secondary male characteristics exhibited in the look and shape of the bird.

Necropsy – post-mortem examination.

Newcastle Disease – highly infectious viral disease leading to respiratory problems with high mortality. Not common in waterfowl.

Nuptial moult – happens to males at the end of the winter, taking them from drab camouflaged plumage to smart colourful feathers.

Oil gland (uropygial or preen gland) – located at the base of the tail, producing water-repellent oil for the feathers.

Pasteurellosis (Duck Cholera) – respiratory infection spread by rodents and wild birds.

Pedigree – a 'family tree' showing the ancestry of a registered, pure-bred animal.

pH – value that indicates the acidity or alkalinity.

Phenotype – observable physical characteristics of an individual.

Pin feathers – new feathers just emerging from the skin.

Pinioning – removing the wing tip of day-old birds of non-native species to ensure they cannot escape into the wild.

Plucking – removing the feathers from a carcase to make it fit to eat.

Pneumonia – an inflammation of the lungs, caused by a bacterial or viral infection.

Pour-on – a chemical preparation for control of internal/external parasites that is applied to the skin and is gradually absorbed; an alternative to injectable treatments.

Predator – an animal that lives by hunting, killing and eating other species.

Preen – to smooth, arrange and clean feathers with the beak or bill.

Probiotic – living organism used to manipulate fermentation in the gut.

Progeny – offspring of an animal.

Prolapse – interior organ(s) pushed outside of the body cavity (e.g. oviduct).

Prolific – highly productive egg layer.

Pullet – young female duck in her first year of egg production.

Pure-bred – not crossed with another breed.

Quarantine – confine and keep an animal away from the rest of the flock to prevent the spread of disease.

Roached back – deformed, arched, or humped back.

Rotational grazing – organized system of moving stock from one grazing unit to another.

Roughage – high-fibre feed that is low in both digestible nutrients and energy.

Roundworm – unsegmented parasitic worms with elongated rounded bodies that are pointed at both ends.

Scouring – *see* 'diarrhoea'.

Sex change – adult birds changing from female to male, can be temporary or a permanent change.

Sex feathers – the curled feathers at the centre of a drake's tail.

Sharps – needles, syringes, scalpel blades, and anything else that can puncture the skin.

Sire – father.

Sport – a mutant such as a white sport in a bird of a breed that usually never has white markings.

Sticky eye – eye infection.

Stocking density/stocking rate – the relationship between the number of animals and an area of land.

Straw – the stems of cereals like wheat, barley or oats that are cut and baled and used for bedding.

Strip-grazing – controlling grazing by confining animals to specific areas of land (often using electric fencing) for short periods of time before moving them on to fresh ground.

Stun – to render unconscious, particularly prior to slaughter.

Subcutaneous injection – given under the skin, but not into the muscle; sometimes shortened to sub-Q or SQ.

Supplement – feed/minerals designed to provide nutrients deficient in the animal's main diet.

Sustainable farming – approach that uses on-farm resources efficiently and reduces demands on the environment.

Syrinx – voice box in a bird's trachea.

Tapeworm – ribbon-like parasitic flatworms found in the intestines.

Trio – breeding unit of birds consisting of two females and one male.

Urea – main end-product of protein metabolism in animals.

Vaccine – injection given to improve resistance /prevent disease.

Vent – opening from which the egg or faeces is ejected.

Wet feather – lack of oil on plumage.

Wing clipping – temporary measure clipping off primary feathers on one wing to stop birds flying.

Withdrawal period – after treatment with a medical product, the amount of time that must be allowed to elapse before meat or eggs are allowed into the human food chain.

Wormer – commonly used term for anthelmintic, medication for killing intestinal worms.

Worms – intestinal parasitic worms.

Zero grazing – a system of growing fodder but not allowing livestock to graze it directly; instead, the crop is cut and taken to the animals.

Zoonosis – a disease or ailment that is zoonotic; one that normally exists in animals but can be passed to humans.

Further Information

British Waterfowl Association
A registered charity dedicated to education about waterfowl and their conservation, as well as to raising the standards of keeping and breeding wildfowl, domestic ducks and geese in captivity.
T: 01531 671250
E: secretary@waterfowl.org.uk
W: www.waterfowl.org.uk

Domestic Waterfowl Club of Great Britain
Ros King, Limetree Cottages, Brightwalton, Newbury, Berkshire, RG20 7BZ
E: rosking@freenetname.co.uk
W: www.domestic-waterfowl.co.uk

Poultry Club of Great Britain
Chattlehope House, Catcleugh, Newcastle upon Tyne, NE19 1TY
T: 01830 520856
E: info@poultryclub.org
W: www.poultryclub.org

Indian Runner Duck Club
Christine Ashton (Secretary), Red House, Hope, Welshpool, Powys, SY21 8JD
E: info@runnerduck.net
W: www.runnerduck.net

International Wild Waterfowl Association
E: info@wildwaterfowl.org
W: www.wildwaterfowl.org

Poultry Club of Wales
E: poultryclubofwales@outlook.com
W: www.poultryclubofwales.co.uk

Poultry Keeper
E: admin@poultrykeeper.com
W: poultrykeeper.com

British Poultry Council
T: 020 3544 1675
E: info@britishpoultry.org.uk
W: britishpoultry.org.uk
Twitter: @britishpoultry

British Goose Producers (part of the British Poultry Council)
T: 020 3544 1675
E: info@britishpoultry.org.uk
W: britishgoose.org.uk

British Call Duck Club
T: 01437 721433
E: alan.j.davies@talktalk.net
W: www.britishcallduckclub.org.uk/

Goose Club
T: 01437 563309
E: contact@gooseclub.org.uk
W: www.gooseclub.org.uk/

GOVERNMENT AND OTHER HELPFUL ORGANIZATIONS

Department for Environment, Food and Rural Affairs (DEFRA) and Animal and Plant Health Agency (APHA)
In England T: 03000 200 301
In Wales T: 0300 303 8268
In Scotland Call the relevant field office:
Ayr T: 03000 600703
Galashiels T: 03000 600711
Inverness T: 03000 600709
Inverurie T: 03000 600708
Perth T: 03000 600704
E: APHA.Scotland@apha.gov.uk
In Northern Ireland Department of Agriculture, Environment and Rural Affairs (DAERA) www.

daera-ni.gov.uk/contact T: 0300 200 7843 E: daera.helpline@daera-ni.gov.uk

Food Standards Agency (FSA) www.food.gov.uk/
In England: T: 0330 332 7149
E: helpline@food.gov.uk
In Wales E: walesadminteam@food.gov.uk
In Northern Ireland E: infosani@food.gov.uk
In Scotland Food Standards Scotland (FSS)
www.foodstandards.gov.sco T: 01224 285100
E: enquiries@fss.scot

Farming and Wildlife Advisory Group
(FWAG)
E: info@fwag.org.uk
W: www.fwag.org.uk

Humane Slaughter Association
T: 01582 831919
E: info@hsa.org.uk
W: www.hsa.org.uk

Moredun Research Institute
T: 0131 445 5111
E: info@moredun.org.uk
W: www.moredun.org.uk

National Animal Disease Information Service
(NADIS)
T: 07771 190823
E: contact@nadis.org.uk
W: www.nadis.org.uk

National Fallen Stock Company
T: 01335 320014
E: member@nfsco.co.uk
W: www.nfsco.co.uk

Rare Breeds Survival Trust (RBST)
T: 024 7669 6551
E: enquiries@rbst.org.uk
W: www.rbst.org.uk

National Office of Animal Health (NOAH)
T: 0208 3673131
E: noah@noah.co.uk
W: www.noahcompendium.co.uk

EQUIPMENT

There are many sources of poultry equipment, from local agricultural merchants to on-line stores. The ones listed below keep items of particular interest to the duck and goose keeper.

Animal Arks
(housing)
T: 01579 382743
E: arks@animalarks.co.uk
W: www.animalarks.co.uk

Brinsea
(incubators, hatchers and candlers)
T: 0345 226 0120
E: sales@brinsea.co.uk
W: https://brinsea.co.uk/

Domestic Fowl Trust
(housing, birds and more)
T: 01789 850046
E: dft@domesticfowltrust.co.uk
W: www.domesticfowltrust.co.uk

Interhatch
T: 01246 264646
E: sales@interhatch.com
W: www.interhatch.com

Jim Vyse Arks
(housing)
T: 01264 356753
E: jimvysearks@aol.com
W: www.jimvysearks.co.uk

Leisure Heating
(for ceramic heat emitters)
T: 0115 937 2727
W: www.leisureheating.co.uk

Little Fields Farm
T: 01455 393 000
E: sales@littlefieldsfarm.com
W: www.littlefieldsfarm.com

Little Peckers
(goose and duck feed)
E: support@petspectrumgroup.co.uk
W: www.littlepeckers.co.uk

Osprey Ltd
(feeders and drinkers)
T: 01588 673821
E: bec@osprey-plastics.co.uk
W: www.bec.uk.com

Sedgbeer Poultry Processing Equipment
T: 01761 420058
E: info@sedgbeer.co.uk
W: www.sedgbeer.co.uk

Solway Recycling
(housing)
T: 01387 730 666
E: info@solwayrecycling.co.uk
W: www.solwayrecycling.co.uk

BOOKS AND BOOKLETS

Ashton, Chris *Keeping Geese: Breeds and Management* (The Crowood Press Ltd, 2012).
Ashton, Chris and Mike (eds) *British Waterfowl Standards* (British Waterfowl Association, 2008).
Ashton, C. and M. *The Indian Runner Duck* (Feathered World, 2002).
Ekarius, *Carol Storey's Illustrated Guide to Poultry Breeds* (Storey Publishing, 2007).
Gooders, John and Boyer, Trevor *The Pocket Guide to Ducks of Britain and Northern Hemisphere* (Parkgate Books, 1998).

Holderread, *Dave Storey's Guide to Raising Ducks* (Storey Publishing, 2001).
Humane Slaughter Association Practical Slaughter or Poultry Guide for the Smallholder and Small Scale Producer.
Roberts, Michael and Victoria *Domestic Duck and Geese in Colour* (The Domestic Fowl Trust, 1986).
Roberts, Michael *Ducks and Geese at Home* (The Domestic Fowl Trust, 1996).
Roberts, Michael *Poultry and Waterfowl Problems* (Domestic Fowl Research, 1998).
Soanes, Barbara *Keeping Domestic Geese* (Blandford, 1992).
Stromberg, Loyl *Sexing All Fowl, Baby Chicks, Game Birds, Cage Birds* (Stromberg Publishing Company, 2002).
Thear, Kate and Dr Fraser, Alistair (eds) *The Complete Book of Raising Livestock and Poultry* (Pan, 1988).
The Goose: History, Folklore, Ancient Recipes (Könemann Verlagsgesellschaft MBH, 1998).
Tyne, Tim and Dot *Viable Self-Sufficiency* (Home Farmer, 2016).

DUCK- AND GOOSE-KEEPING COURSES

Terras Farm (Duck incubation and hatching courses)
Tanya and Roger Olver,
T: 01726 882 383
E: olver@terrasfarm.co.uk
W: www.terrasfarm.co.uk

There are very few goose or duck specific courses, but there are plenty of chicken-keeping courses and introduction to smallholding courses (including our own www.smallholdertraining.co.uk), which may cover the basics of keeping ducks and geese.

Index